T0223801

Embedded Systems Interfacing for Engineers using the Freescale HCS08 Microcontroller I: Assembly Language Programming

Synthesis Lectures on Digital Circuits and Systems

Editor
Mitchell A. Thornton, *Southern Methodist University*

iv

Atmel AVR Microcontroller Primer: Programming and Interfacing
Steven F. Barrett, Daniel J. Pack
2007

Pragmatic Logic
William J. Eccles
2007

PSpice for Filters and Transmission Lines
Paul Tobin
2007

PSpice for Digital Signal Processing
Paul Tobin
2007

PSpice for Analog Communications Engineering
Paul Tobin
2007

PSpice for Digital Communications Engineering
Paul Tobin
2007

PSpice for Circuit Theory and Electronic Devices
Paul Tobin
2007

Pragmatic Circuits: DC and Time Domain
William J. Eccles
2006

Pragmatic Circuits: Frequency Domain
William J. Eccles
2006

Pragmatic Circuits: Signals and Filters
William J. Eccles
2006

High-Speed Digital System Design
Justin Davis
2006

Introduction to Logic Synthesis using Verilog HDL
Robert B.Reese, Mitchell A.Thornton
2006

Microcontrollers Fundamentals for Engineers and Scientists
Steven F. Barrett, Daniel J. Pack
2006

© Springer Nature Switzerland AG 2022

Reprint of original edition © Morgan & Claypool 2009

All rights reserved. No part of this publication may be reproduced, stored in a retrieval system, or transmitted in any form or by any means—electronic, mechanical, photocopy, recording, or any other except for brief quotations in printed reviews, without the prior permission of the publisher.

Embedded Systems Interfacing for Engineers using the Freescale HCS08 Microcontroller I:
Assembly Language Programming
Douglas H. Summerville

ISBN: 978-3-031-79796-5 paperback
ISBN: 978-3-031-79797-2 ebook

DOI 10.1007/978-3-031-79797-2

A Publication in the Springer series
SYNTHESIS LECTURES ON DIGITAL CIRCUITS AND SYSTEMS

Lecture #20
Series Editor: Mitchell A. Thornton, *Southern Methodist University*

Series ISSN
Synthesis Lectures on Digital Circuits and Systems
Print 1930-1743   Electronic 1930-1751

Embedded Systems Interfacing for Engineers using the Freescale HCS08 Microcontroller I: Assembly Language Programming

Douglas H. Summerville

State University of New York at Binghamton

SYNTHESIS LECTURES ON DIGITAL CIRCUITS AND SYSTEMS #20

ABSTRACT

The vast majority of computers in use today are encapsulated within other systems. In contrast to general-purpose computers that run an endless selection of software, these embedded computers are often programmed for a very specific, low-level and often mundane purpose. Low-end microcontrollers, costing as little as one dollar, are often employed by engineers in designs that utilize only a small fraction of the processing capability of the device because it is either more cost-effective than selecting an application-specific part or because programmability offers custom functionality not otherwise available. *Embedded Systems Interfacing for Engineers using the Freescale HCS08 Microcontroller* is a two-part book intended to provide an introduction to hardware and software interfacing for engineers. Building from a comprehensive introduction of fundamental computing concepts, the book suitable for a first course in computer organization for electrical or computer engineering students with a minimal background in digital logic and programming. In addition, this book can be valuable as a reference for engineers new to the Freescale HCS08 family of microcontrollers. The HCS08 processor architecture used in the book is relatively simple to learn, powerful enough to apply towards a wide-range of interfacing tasks, and accommodates breadboard prototyping in a laboratory using freely available and low-cost tools.

In *Part I: Assembly Language Programming*, the programmer's model of the HSC08 family of processors is introduced. This part leads the reader from basic concepts up to implementing basic software control structures in assembly language. Instead of focusing on large-scale programs, the emphasis is on implementing small algorithms necessary to accomplish some of the more common tasks expected in small embedded systems. The first part prepares the reader with the programming skills necessary to write device drivers in and perform basic input/output processing Part II, whose emphasis is on hardware interfacing concepts.

KEYWORDS

microcontrollers, embedded computers, computer engineering, digital systems, Freescale HCS08, assembly language programming

Contents

Acknowledgments

Most of all, I thank Erika, Nicholas, Daniel, and Stephanie for their support, encouragement, and for tolerating all the weekends and evenings that I spent working on this book instead of being with them. This book is dedicated to you. I also thank all the students who debugged the early versions of this book and whose curiosity, questions, and comments helped to shape the presentation of this material.

Douglas H. Summerville
June 2009

CHAPTER 1

Introduction to Microcomputer Organization

This book is about computer organization. The primary focus is on the design of systems that use computers as embedded components, rather than on the design of traditional general-purpose computer systems. One can study computer organization in the context of either of these computer types and have the same basic understanding of computer concepts: operation of the central processing unit, knowledge of fundamental data types, manipulation of data, assembly language programming, interrupts and input/output. The requirements of each type of computer, however, necessitate focusing on specific topics that, while being important to study in one domain, may not be as important in the other. For example, embedded computers generally have limited hardware resources, requiring the system designer to consider many of the low-level details of the computer to ensure optimal application of these resources to the specific application. Concepts such as memory caches, operating systems and file input/output generally do not apply. On the other hand, traditional computers have become so fast and complex as to require high levels of abstraction in order to perform even the simplest task; the programmer neither has nor requires access to low-level system resources. Concepts such as hardware interfacing and low-level device manipulation need to be overlooked. The majority of electrical and computer engineers will encounter the need to design computers as embedded design components rather than design traditional general purpose computers; thus, the focus of this book is on the hardware and software interfacing concepts necessary to use embedded computers as components in larger systems.

The Freescale HCS08 CPU architecture is the focus of this book because it is a small, easy to learn processor architecture that is inexpensive and easily used by hobbyist and students in a traditional breadboard environment with freely available tools. The processor is also powerful enough to accommodate the requirements of many practical embedded systems.

This chapter introduces some fundamental concepts of computer organization, defines common computer engineering terms and introduces basic data representation and manipulation.

1.1 DATA AND INFORMATION

Two computing terms often confused are data and information. This confusion is understandable, given that there are many discipline-specific definitions of these terms that, although suitable in particular fields, are inappropriate or otherwise useless in others. A recurring theme among these definitions is that processing of data results in information. A convenient definition suitable for computing in general is *information* is "data with context." By itself, data has no meaning. How a computer program accesses and uses a data value assigns meaning to it. By combining and processing a number of data values, a program extracts information. In other words, how a program manipulates the data provides the context necessary to define its value. In combination with other values, information is produced.

Just as data can be processed to extract information, information must be converted into structured data to be stored in digital form. For example, music (information) is converted into values that are encoded into bytes to be stored on an audio CD. A computer can run a program to process these bytes back into an audio (information) signal that can be played back through a pair of speakers.

Even a single datum, in isolation, has no real meaning. For example, if a program were to input the value 48 at a particular location in memory, it can be interpreted as having many values, including (but not limited to)

- the unsigned value 48

- the ASCII character '0'

- the packed binary-coded decimal number 30

This datum could be a value used by a program to represent temperature; the program might manipulate the value as an unsigned integer and combine it with the text data "The temperature outside, in degrees Fahrenheit, is" to produce information about how warm it is outside.

1.2   ORGANIZATION OF DIGITAL COMPUTERS

A *computer* is a device that is capable of processing data under control of a program. A *computer program* is a sequence of instructions from the computer's instruction set that implements an *algorithm*, which is a list of steps for accomplishing a specific task. Today, the term computer is generally used to mean a "stored-program binary digital electronic computer." "Stored-program" means that the computer has the flexibility to execute a variety of programs stored in its memory; that is, its functionality can be tailored to application needs. In contrast, a fixed-program computing device, like a simple pocket calculator or digital wristwatch, has dedicated functionality designed into the hardware. "Binary digital electronic" means that data is stored and manipulated in electronic form using only two states: 1 and 0. In this context, the smallest amount of data is the bit (**b**inary dig**it**), whose value can be 0 or 1.

There are many types of computers in use today, from low-end personal digital assistants (PDAs) to supercomputers costing millions of dollars. Most people associate the word computer with a traditional desktop or personal computer. These *general-purpose computers* can be programmed to perform a wide variety of tasks such as processing email, audio and video playback, word-processing, Internet browsing, and gaming. To provide such flexibility, general-purpose computers have an *operating system* that manages system resources and provides a hardware-independent foundation on which complex applications can be built with relative ease.

By far, the majority of computers in use today are hidden inside of products that do not resemble "traditional" computers. Examples include cell phones, watches, automobiles (which typically have several), MP3 players, household appliances and consumer electronics. The computers in these devices are called *embedded computers* because they are completely encapsulated inside the device they control, to the extent that most users are not even aware that they are interacting with a computer at all. These computers are generally application-specific, as opposed to general purpose, because they cannot be easily tailored to perform other applications.

A basic computer system can be divided into 3 major functional blocks: central processing unit (CPU), memory and Input/Output (I/O). As illustrated in the Figure 1.1, these components are interconnected by a system bus, which is a set of signal lines (wires) over which the components exchange data. Although some computers have more complex structures than that shown in the figure, such as separate busses for memory and input/output, the simple model shown provides sufficient detail to understand the basic operation of a computer.

1.2.1 CENTRAL PROCESSING UNIT

The CPU is both the workhorse of the computer as well as the supervisor. It is responsible for stepping through the program and performing the specified arithmetic, logical and data transfer instructions listed. The CPU controls the memory and input/output units to support the execution of operations specified by the program instructions. The memory and input/output units are, in this respect, passive – they function under control of the CPU. The types, formats and sizes of data supported as well as the set of instructions included differentiate one CPU from another. The *programmer's model* of the CPU is everything a programmer needs to know about the CPU in order to program it. Low-level hardware details about how the processor is implemented are generally not part of the programmer's model.

The CPU performs all operations under control of a program. A program consists of a sequence of instructions stored in memory. Each CPU has a fixed set of instructions that it can understand. Some CPUs use only very simple, regular instructions that perform small operations on data, such as add and subtract, which can be completed in small fixed-number of clock cycles. These are referred to as Reduced Instruction Set Computers (RISC, pronounced 'risk'). Other CPUs use more complex instruction sets, using irregular instructions having differing complexities and that take a variable amount of time to complete. These are referred to as Complex Instruction Set Computers (CISC, pronounced 'sisk'). In essence, a CISC instruction can be thought of as performing the equivalent of several RISC instructions. The popular Intel processors used in many personal computers and workstations are CISC.

A CPU has a small number of *registers* that are used to hold data being operated on by the CPU as well as to maintain program state. Access to register data is faster than access to data in memory because the data is already inside the CPU. Some CPUs can operate only on data contained within the CPU registers. In these CPUs, data values must be explicitly *loaded* into CPU registers before they can be used in operations. Arithmetic and logical instructions then operate on data in the

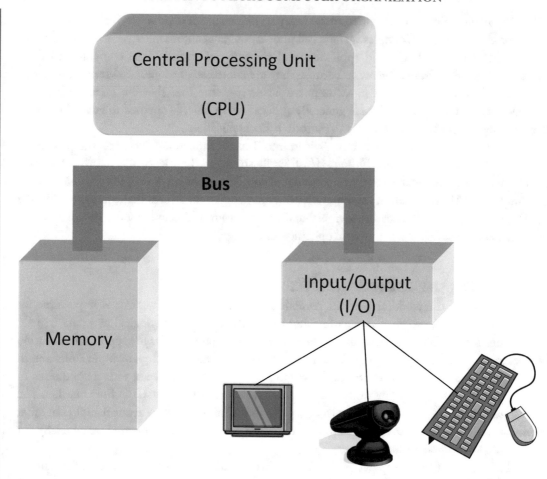

Figure 1.1: Organization of a basic computer system.

registers and place results back into registers. The results must be explicitly copied (*stored*) into main memory if they are to be retained by the program. A CPU using this model of computation is said to have a load-store architecture, and most RISC CPUs have load-store architectures. In contrast, some CPUs have arithmetic and logical instructions that can operate directly on data values contained in main memory and store results directly back into memory (bypassing the registers). These are referred to as memory-memory or register-memory architectures. Here, registers provide temporary storage for intermediate results computed by the CPU.

CPU instructions, like all data in memory, must be stored in binary form. The binary form of a program, which is understandable by the CPU, is called *machine code*. An example of a machine code instruction from the Freescale HCS08 CPU instruction set is shown in Figure 1.2. This machine

$$\underbrace{10101011}_{\text{Opcode (ADD)}} \quad \underbrace{00000011}_{\text{Operand (3)}}$$

Figure 1.2: Example machine code instruction.

code represents the instruction that tells the CPU to add the constant value 3 to the contents of register A: or A=A+3. The *opcode* is the part of the machine code that represents the operation – in this example, the operation is "add constant to register A." The *operand(s)* represent the data to be used in the instruction. In this example, there are three operands: A is implicitly specified as the destination of the addition, as well as the first addend, by the opcode; the constant 3 is explicitly listed as the second addend of the addition. Together, the opcode along with its listed operands make up a machine code instruction. Thus, this instruction uses 16 bits (or two bytes) for its machine code. CPUs often have a variety of methods to specify operands. Operands can be constants contained in the machine code (called immediates), as in the example above, or can be values stored in memory, or can be values stored in registers in the CPU.

It is tedious for humans to work directly with machine code. Thus, a common textual form of the instruction set of the CPU is defined by the manufacturer; this textual form is called an *assembly language*. The standard assembly language corresponding to the machine code in Figure 1.2 is ADD #3. Each opcode is assigned a mnemonic (pronounced "new-monic"), which is a textual representation of the opcode that is easier to remember. Example mnemonics include ADD, SUB, and INC, for which you can likely guess the operation specified without knowing the instruction set (try to do that with machine code!). The opcode mnemonics are easier to remember than binary machine code values. Each assembly language instruction converts directly to a single machine code instruction; thus, assembly language is simply another form of the machine code. The CPU, however, cannot interpret assembly language directly. A program, called an *assembler*, takes the assembly language from a text file and converts it to machine code. If the assembler runs on the same CPU that it assembles code for, then it is a native assembler; otherwise, it is a cross-assembler. For example, Freescale HCS08 programs are generally assembled on a standard PC using a cross assembler.

For large programs on complex computers, assembly language can be tedious to work with. In addition, assembly language written for one computer type cannot run on another – it must be carefully translated from one machine code into another, an extremely tedious task that is not unlike translating one spoken language to another. High-level languages (HLLs) such as C, BASIC, Java, and others have been developed to provide more powerful languages that are computer independent. Each line in a HLL might correspond to many assembly language instructions. A program written in a HLL needs to be compiled before it can be executed. A *compiler* is a program that interprets the HLL program and converts it into the assembly language (or machine code) of just about any CPU. HLLs have a higher level of abstraction than assembly language, meaning they abstract the details of the computer they are running on from the programmer. This abstraction is what allows platform independence.

During its operation the CPU repeatedly performs the following basic sequence of steps, which is called instruction sequencing. Depending on the CPU or instruction type, the actual sequence might be slightly different. The steps are

1. **Instruction Fetch**: The CPU reads the machine code of the next instruction from memory. The CPU has a special register, the *Program Counter*, which holds the address of the next instruction to be executed. The program counter is updated to point to the next instruction after each instruction.

2. **Decode**: The machine code is decoded (interpreted) by the CPU into an operation and operands.

3. **Execute**: The operation specified by the opcode is performed on the data specified as operands.

4. **Store Results**: if required, results are stored to memory or registers.

This sequence is repeated millions or billions of times per second on modern CPUs. There are circumstances, however, that require instruction sequencing to be altered or suspended. Such circumstances arise from two types of situations. The first is when the instruction sequencing itself encounters a problem. This type of error is often referred to as an *exception*. Examples include when the CPU fetches an opcode that does not represent a valid instruction or when a divide instruction encounters a divisor of zero. How these errors are dealt with depends on the specific needs of the system or program. The second type of situation that alters instruction sequencing has to do with input/output. Certain events can happen unpredictably, such a keyboard keystroke, a packet arrival on a network interface and triggering an alarm on a security system. Most CPUs have an *interrupt mechanism*, through which normal instruction sequencing is suspended to allow the CPU to deal with these events; sequencing is later resumed once the event has been dealt with. The specific mechanisms for handling interrupts vary for each CPU, although they are usually designed to handle both interrupts and exceptions. The system designer can create a special program segment, called an *interrupt service routine*, to deal with each event as it occurs. The alternative to having an interrupt mechanism is for the programmer to include code in every program that repeatedly checks for these events, but this complicates programming and wastes computing time.

1.2.2 MEMORY

Memory in a computer has a similar purpose as biological memory – it provides the ability to retain data. The primary use of memory in a computer system is to hold programs and their variables. A computer memory can be logically thought of as a number of fixed-sized locations, each of which holds a program instruction or data value. Each memory location is assigned a unique number, called its *address*, which the CPU uses to refer to the location explicitly. A CPU accesses both data and instructions by referring to their memory addresses.

The maximum length of physical memory that a CPU can directly access is limited by the address size it uses. The *address space* is defined as the number of memory locations that the CPU can access and is calculated as

$$\text{address space} = 2^{n_a}$$

where n_a is the address size, in bits, used by the CPU. For example, a processor that uses 16 bit memory addresses can access up to 65536 memory locations, which are numbered from 0 to 65535.

Each memory address holds a fixed number of bits. The maximum width of the memory that can be used is limited by the number of bits that the CPU can transfer to/from memory at a given time. This typically matches the number of bits that can be processed by the CPU with a single instruction, often called the word size. Typical word sizes are 8 or 16 bits for embedded computers and 32 or 64 bits for general-purpose desktop computers and workstations. If a computer has a word size of 16, 32, or 64 bits, it still has the option of uniquely addressing to each memory byte; in that case, the computer is said to have a byte-addressable memory. If each memory address corresponds to a unit of data greater than 1 byte, it is called word-addressable.

The word "memory" is typically used to refer to only the primary storage in the computer system – that which can be directly accessed by the CPU. The term *storage*, or *secondary storage*, is used to describe devices such as disks and tapes, from which data is indirectly accessed by the CPU. The CPU goes through another hardware device, by way of the input/output unit, to access secondary storage. This is why a computer can access an almost unlimited amount of secondary storage, even with a small address space. Some CPUs use separate address spaces for data and programs: a *program address space* and a *data address space*. CPUs that have this separation of data and programs are said to have a *Harvard Architecture*, while CPUs with a combined data/address space are said to have a *von Neumann* (pronounced "von noi-man") architecture.

A memory read operation occurs when the CPU needs to obtain the contents of a memory location; the direction of data transfer is from memory to CPU. A memory write occurs when the CPU needs to change the contents of a memory location; in this case, the direction of data transfer is from CPU to memory.

There are many types of memories available, but all can be divided into two categories: volatile and non-volatile memories. One can think of volatile memory as being the computers short-term memory, while long-term memory capability is provided by nonvolatile memory (and secondary storage). A *volatile memory* loses its contents when the power is turned off. The contents of a volatile memory are usually undefined when powered on. A *nonvolatile memory* retains its contents even when the power is turned off for a long period of time – when the memory is powered on, the data values are the same as they were when it was last powered off. While a computer could just use non-volatile memory, the technologies used in the two types of devices makes volatile storage faster and cheaper, making it more suitable for use as the primary writeable memory for holding programs and data. Nonvolatile memory is used for fixed programs, constant data and configuration settings. For example, when a computer is first powered on, it needs a small program to help initialize the computer system to begin normal operation: this *bootstrapping program* is an example of a fixed program that is typically stored in non-volatile memory. In embedded systems with no secondary storage such as disk drives, non-volatile memory is the primary option available for storing programs and constant data. This embedded software is called *firmware*.

The primary type of volatile memory used as writeable memory in computers is called Random Access Memory, or RAM (pronounced as one word "ram"). RAM comes in a variety of types; logically, however, each has the ability to access memory locations in any order, hence the use of the word random in the name. In primary use today are two types of RAM technologies: static RAM (SRAM, pronounced "ess-ram") and dynamic RAM or DRAM (pronounced "dee-ram"). Most RAM types are volatile; there are a few non-volatile RAMs (or NOVRAMs), which use either battery backup or backup to nonvolatile memory when powered off to retain their contents. However, these devices tend to be small and expensive per bit and therefore not used as the primary writeable memory in computers.

The basic nonvolatile memory is called a read only memory or ROM (pronounced "rom" rhyming with bomb). Devices that have data fixed in them by the semiconductor manufacturer at the time of manufacture are called *mask ROMs*; other devices can be written by the customer only once and are referred to as Programmable ROMs or PROMs (pronounced as one word "prom," like the high-school dance). These devices are also called one-time-programmable or *OTP ROMs*. OTP ROM chips generally require a standalone programmer (or "burner") to write the permanent memory data to them. Although ROMs do not provide the flexibility to change their data contents after initial programming, their high density, low power and low cost make them attractive for applications in which the is no need to update or modify stored data, or where security concerns might require that data not be changed.

Rewriteable nonvolatile memory technologies continually improve in density, cost and ease of reprogramming. Reprogramming involves erasing and then writing the new contents into the device. Erasable PROM or EPROM (pronounced "ee-prom") is a type of rewriteable ROM that is erased by exposing the chip to ultraviolet light through a quartz window, which is covered with opaque tape during normal operation to prevent accidental exposure. Electrically erasable programmable ROM, or EEPROM for short (pronounced "ee-ee-prom"), evolved from EPROM but can be erased electrically, which is more convenient and makes it possible to perform both programming and erasing operations in-circuit (the device does not have to be removed from the circuit and placed into a separate programmer). This in-system-programmability (ISP) has led to widespread use of EEPROM devices for holding configuration data in both embedded and traditional computer systems. Flash memory devices are similar to EEPROMs but have been developed as replacements for traditional magnetic media such as floppy disks. Unlike EEPROM, which can be reprogrammed a single byte at a time, flash memory is reprogrammed a block at a time. Both Flash memory and EEPROM are limited in the number of times that they can be reprogrammed (10,000 or more programming cycles is common) and both have much slower read access and write (programming) times than RAM.

The choice of type of nonvolatile memory to use is based on the application for which it is needed. Some guidelines are summarized in Table 1.1. It should be noted that EEPROM and Flash technologies are rapidly changing and the distinction between them is blurring. Flash devices that emulate the ability to modify individual data bytes in memory are emerging as well as devices that allow protecting memory contents from unauthorized modifications (to protect intellectual property, for example).

Table 1.1: Comparison of nonvolatile memory technologies and guidelines for their selection.

Memory Type	Characteristics	When to Use
Mask ROM	Data programmed into device during manufacture; it cannot be changed.	Selected when large quantities of identical ROMs are required and contents never require updating. Not a viable option if a small number of ROM devices are needed.
PROM (OTP ROM)	Programmed only once; contents cannot be changed.	Selected when small quantities of a ROM are needed and the contents do not require updating. Semiconductor maker does not need to preprogram contents.
EPROM	Can be reprogrammed, but ROM chip must be removed, erased, and reprogrammed	Not presently in widespread use because electrically erasable devices provide the same reprogramming capability.
EEPROM	Individual bytes can be electrically erased and reprogrammed in circuit.	EEPROM is suitable for use where modification of individual data bytes is necessary and memory capacity is moderate (e.g., for configuration data).
Flash	Can be electrically erased and reprogrammed, but only in blocks of multiple bytes.	Suitable for applications where updates require changing large sections of data and/or a large amount of space is required (e.g., firmware, bootstrapping programs, bulk data).

1.2.3   INPUT AND OUTPUT

The Input/Output unit is a collection of hardware interfaces used by the CPU to interact with peripheral devices and other computers. A *peripheral* device is a device that extends the functionality of the computer. Examples of peripheral devices include DVD drives, keyboards, video cards, microphones, printers and cameras. An *interface* is a well-defined specification for communication between two devices, defining the mechanical, electronic and information standards that allow the devices to exchange data. The term port is sometimes used interchangeably with interface, but generally refers to the computer side of the interface. Thus, one talks of interfacing a peripheral to a specific computer port. Examples of standard interfaces include serial (RS232, later renamed EIA232), parallel (IEEE 1284 standard), Universal Serial Bus (USB), PS/2 (developed by IBM and still in use on some personal computers as the mouse port) as well as many others.

1.2.4   BUS

The bus consists of the electrical connections over which the main components of the computer exchange data. Usually, the CPU is the sole master of the bus, meaning that all bus transfers occur under its control. The other bus components have a more passive role, responding to requests from the CPU. In this simple model, every bus operation involves a transfer of data between the CPU and either memory or I/O. Logically, the bus can be divided into three separate types of signals: address, data and control. The address lines are used to identify the component that is to be the partner in the data exchange with the CPU as well as specify the location of the resource (memory location or I/O register). Typically, a single exchange of data can occur over the bus during a bus cycle (usually the same as a bus clock cycle). Thus, multiple clock cycles can be required for the CPU to complete an operation.

1.2.5   MICROPROCESSORS, MICROCOMPUTERS AND MICROCONTROLLERS

In the early days of computing, CPUs were built from a large number of components, first from bulky discrete devices and later from small scale integrated circuits, which are circuits constructed from miniaturized components implemented on a thin semiconductor substrate (often referred to as microchips or simply chips). As technology improves, more and more functionality can be included on a chip. A CPU implemented on a single integrated circuit is called a *microprocessor* (CPU on a chip). The first microprocessors were introduced in the early 1970s.

A computer built using a microprocessor is a *microcomputer*. The name microcomputer does not imply that the computer is small, only that it has a microprocessor. As the amount of circuitry that could be included on a single integrated circuit increased, more of the functionality of the microcomputer began to be moved on chip with the microprocessor. A single chip containing a microprocessor, memory and input/output devices is called a *microcontroller* (in other words, a microcontroller is a microcomputer on a chip).

Because there are a wide variety of input/output devices and memory configurations that could be used, microcontrollers generally exist in families that share a common CPU but have different input/output and memory configurations. When choosing a microcontroller for an application, one generally selects the family first and then chooses a member from the family that provides the closest match to the input/output and memory requirements of the application.

1.3   DATA REPRESENTATION

It is important to remember that computers operate on data in binary form. A binary value stored in memory has no meaning until it is manipulated by a program; how the program uses the data provides the context necessary to assign meaning to it. CPUs have instructions that operate on data using predefined data formats or codes – a code is a rule for converting information into another form. A CPU instruction that manipulates unsigned integer operands would attribute a different value to the binary operand 10000001 than an instruction manipulating signed operands. A CPU can manipulate data directly using a few standard codes; for other codes, the programmer has the responsibility of programming the algorithms for encoding, decoding and manipulating the data.

In computing, several bases are commonly used to represent numbers. Base 2 (binary), 8 (octal), 10 (decimal) and 16 (hexadecimal) are commonly used. When it is not clear, the base of the number should be specified by subscripting it with the numerical value of the base. For example, to avoid confusion the binary number 100011101 can be written as 100011101_2.

1.3.1   CODES AND PRECISION

Data is encoded into binary form using a code. The number of bits used to encode the data limits the total number of values that can be represented. For example, a 1 bit code can only encode data into one of two values: 0 or 1. Precision is defined as the number of values or alternatives that can be encoded. With 1 bit, one can encode 2 values or alternatives; with 2 bits, one can encode 4 values. The more values that need to be encoded, the more bits needed to store the data. Thus, for an n-bit code, the precision M is

$$M = 2^n$$

and, if one needs to encode M different values or alternatives, the number 'n' of bits required is

$$n = \lceil \log_2 M \rceil$$

The ceiling notation is required because data must be encoded using a whole number of bits, and we cannot use fewer bits that required. Note that the unit of n is "bits," represented by lowercase 'b'; uppercase 'B' represents units of bytes.

Since data can only be encoded using whole numbers of bits, practical codes have precisions that are powers of two (2,4,8,16,32,…). Thus, it is generally not necessary to use a calculator to compute precision. One can simply find the smallest power of two that equals or exceeds the required precision.

Example 1.1.   A programmer needs to store 3 values in memory to represent a date: the month, day, and year. What is the minimum number of bits required to store each value?
Solution: There are 12 alternatives for the month; therefore $\lceil \log_2 12 \rceil = 4b$ are required to store the value of the month. Notice that no code has been specified – we have not defined what binary value will be used to represent January, February, March, etc. There are at most 31 alternatives for the day of the month; therefore $\lceil \log_2 31 \rceil = 5b$ are required to encode the day of the month. Assuming the value of the year is from 0000–9999, there are 10000 alternatives requiring 14 bits.

Example 1.2.   To encode the hour of the day using twenty four hour time requires 5 bits. Why?
Solution: 4 bits gives a precision of 16, and 5 gives a precision of 32. Thus, 5 is the smallest **number** of bits that gives enough precision.

Because precision is limited to powers of two, there will be instances when precision exceeds the number of alternatives that need to be represented; thus, there will be undefined values for the code. In the example above, we can assign values to the month in many ways: three are shown in Table 1.2. Four bits can encode 16 values and there are 12 alternatives (months) to encode. Any 4-bit code chosen to represent the month will have four unused codewords representing invalid values for the month. Programmers often use these unused codewords as status; for example, setting the value of the month to 15 can be used to indicate that no data is stored.

As a practical matter, of the three alternative codes shown above to represent the month, Code A and Code B have advantages over Code C. Code C is essentially random and does not allow months to be numerically compared or manipulated. Using Codes A or B, the binary values of any two months can be numerically compared and simple arithmetic can be applied (adding 1 to the value of the month will give the value of next month, for example).

1.3.2   PREFIX NOTATION

When specifying a large number, we can use a prefix multiplier to clarify the representation. In engineering notation, one factors out powers of 10 in multiples of 3, for example, and 12,000 is written as 12×10^3. When the number has units, the power of 10 can be replaced with a standard prefix; for example 12,000 grams can be written as 12 kilograms or 12 kg. The standard SI prefixes are shown in Table 1.3(a). Note that "k" is the only SI prefix in lowercase.

In computer engineering practice, prefix multipliers are also used to simplify the representation of large binary values. Historically, the prefix multipliers have followed the same naming convention as in standard engineering notation, but represent different quantities. In computer engineering, the

Table 1.2: Three possible codes for encoding the month of the year.

Codeword (binary value)	Code A	Code B	Code C
0000	January	*Unused*	*Unused*
0001	February	January	*Unused*
0010	March	February	March
0011	April	March	January
0100	May	April	September
0101	June	May	April
0110	July	June	February
0111	August	July	October
1000	September	August	*Unused*
1001	October	September	August
1010	November	October	November
1011	December	November	May
1100	*Unused*	December	July
1101	*Unused*	*Unused*	December
1110	*Unused*	*Unused*	June
1111	*Unused*	*Unused*	*Unused*

Table 1.3: Tables of standard SI and binary prefixes.

(a) SI prefixes.		(b) Binary prefixes.	
SI Prefix	**Equivalent Value**	**Binary Prefix**	**Equivalent Value**
k (kilo)	10^3 (1,000)	Ki (kibi)	2^{10} (1,024)
M (mega)	10^6 (1,000,000)	Mi (mebi)	2^{20} (1,048,576)
G (giga)	10^9 (1,000,000,000)	Gi (gibi)	2^{30} (1,073,741,824)
T (tera)	10^{12} (1,000,000,000,000)	Ti (tebi)	2^{40} (1,099,511,627,776)
P (peta)	10^{15} (1,000,000,000,000,000)	Pi (pebi)	2^{50}
E (exa)	10^{18}	Ei (exbi)	2^{60}

prefixes represent powers of 2 in multiples of 10, resulting in prefix multiplier values close to those used in scientific notation, but not exactly the same. For example, the prefix kilo has been commonly used to represent 2^{10}=1024, instead of 10^3, when the base is understood to be binary. This has led to some confusion, however, when the base is not clearly specified. In 1998, the International Electrotechnical Commission (IEC) defined a new set of computer engineering prefixes, shown in Table 1.3(b). These prefixes follow similar naming convention to the SI prefixes, having a common first syllable but the second syllable being replaced with "bi" (pronounced "bee"). Though not yet widely adopted, their use is becoming increasingly prevalent and has been endorsed by the Institute of Electrical and Electronics Engineers (IEEE).

Example 1.3. Exactly how many bytes is 2 tebibytes?
Solution:
$$2 \text{ TiB} = 2{*}2^{40} \text{ B} = 2{,}199{,}023{,}255{,}552 \text{ B.}$$
Answer: 2,199,023,255,552 B.

Example 1.4. What is the precision of a 12 bit variable, expressed using prefix notation?
Solution: A 12 bit variable can represent 2^{12} different values. To convert to prefix notation, factor out the largest power of 2 that is a multiple of 10. Thus, $2^{12} = 2^2 \times 2^{10} = 4$ Ki.
Answer: the precision is 4 Ki.

Example 1.5. A CPU uses 16 bit addresses to access a byte-addressable memory. How many bytes of memory can be addressed, specified in prefix notation?
Solution:
$$2^{16} \text{ B} = 2^6 * 2^{10} \text{ B} = 64 \text{ KiB}$$
Answer: 64 kibibytes.

1.3.3 HEXADECIMAL REPRESENTATION

While prefix notation can be used to simplify the representation of large numerical values (quantities), it cannot be used to simplify large binary strings. Consider the case when working with 16 bit binary addresses, for example. The address 1000010100001000_2 cannot be approximated because it represents a specific memory location, not just a large value. On the other hand, the address is cumbersome to work with. Working with data in this form is prone to error – copying just one bit incorrectly would indicate a different memory address.

The solution to this problem is to represent the number in a more concise format. In computing, octal (base (8) and hexadecimal (base 16) are commonly used for this purpose. The reason for this is that both bases are themselves powers of 2, allowing direct conversion between each hex (hexadecimal) or octal digit and an integral number of bits. Table 1.4 outlines the correspondence between binary, octal and hexadecimal representations. Each octal digit has a precision of 8, which is equivalent to 3 bits. Similarly, each hex digit is equivalent to 4 bits.

In some cases, the string of bits to be represented in hexadecimal or octal notation does not contain a multiple of 4 or 3 bits, respectively. In these cases, we add leading zeros to the number before converting; the number of bits being represented, however, must be explicitly specified in these cases (when specifying two's complement numbers, sign-extending might be more appropriate than zero-extending, as discussed Section 1.2.2).

Table 1.4: Octal and binary digits and their binary equivalents.

Hexadecimal Digit	Binary Equivalent	Octal Digit	Binary Equivalent
0	0000	0	000
1	0001	1	001
2	0010	2	010
3	0011	3	011
4	0100	4	100
5	0101	5	101
6	0101	6	110
7	0111	7	111
8	1000		
9	1001		
A	1010		
B	1011		
C	1100		
D	1101		
E	1110		
F	1111		

Example 1.6. How would you represent 1010110011111_2 in octal and hexadecimal?
Solution: To convert to octal, group the bits into threes and convert each group of three to its equivalent octal digit

$$\underline{101} \quad \underline{011} \quad \underline{001} \quad \underline{111}$$
$$5 \qquad 3 \qquad 1 \qquad 7$$

To convert to hexadecimal, group the bits into fours and convert each group to its hexadecimal digit

$$\underline{1010} \quad \underline{1100} \quad \underline{1111}$$
$$A \qquad C \qquad F$$

Answer: $1010110011111_2 = 5317_8 = ACF_{16}$.

Example 1.7.   The 7 bit binary string 1110100 can be represented in octal as 164, as shown below.

$$\begin{array}{ccc} 1 & 110 & 100 \\ 1 & 6 & 4 \end{array}$$

However, a 3 digit octal number has more precision ($8^3 = 2^9$) than a 7 bit binary number (2^7). Thus, when representing the binary string as 164_8 it must understood that it is a 7 bit string that is being represented, otherwise this must be explicitly specified (i.e., the 7 bit binary string 164_8). For example, the binary string represented by 211_8 is 010 001 001, but the 7b binary string 211_8 is 10 001 001.

1.3.4   MEMORY MAPS

When it is necessary to illustrate a portion of the contents of memory, a memory map can be used. A memory map is a diagram that represents the contents of memory. Depending on the level of detail that needs to be communicated, two basic types of memory maps can be used, as shown in Figure 1.3. The first is a detailed map showing the contents of individual bytes in memory; each box in a detailed memory map represents one byte in memory; the value of the byte is shown inside the box, while the address is labeled outside the box. The second is a memory map of all or part of the address space, where each box represents an entire region of memory. Address space memory maps are generally used to show how the different types of memory are assigned to the addresses available in the microcomputer. They can also be used to show how programs and data are organized within a region of memory.

A textual form of a detailed memory map is a memory dump or HEX dump. HEX dumps are commonly produced by programming tools, such as debuggers, because they offer a more concise representation of large number of memory values. An example of a typical HEX dump is shown in Figure 1.4. Each line of a HEX dump shows the address of the first byte in the line, followed by the 8 (or 16) consecutive bytes of data starting at that address.

1.4   COMMONLY USED BINARY CODES

As stated earlier, how a CPU manipulates data stored in memory is what provides it with the context necessary to give meaning to it. Each CPU has instructions that process data represented using different codes. The set of codes or data types directly supported by a CPU includes those for which the CPU has instructions to manipulate them directly. Software routines must be provided to manipulate data types not directly supported by the CPU; thus, any CPU can be programmed to

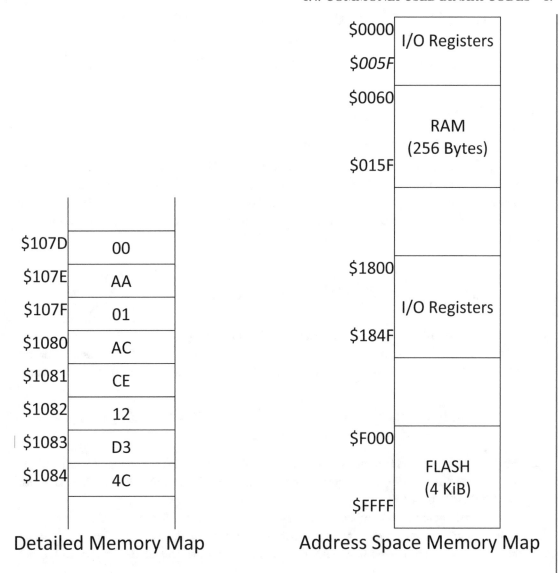

Figure 1.3: Examples of memory maps.

manipulate data using any code desired but are fastest at manipulating data in their "native" codes. Since most operations can be performed faster in hardware, it is advantageous to use supported CPU data types whenever possible.

In this section, commonly supported binary codes are discussed. These include codes to represent character data, unsigned and signed integers and fixed-point numbers.

```
1070    3A 00 00 12 11 AB 88 99
1078    10 38 76 44 30 00 AA 01
1080    AC CE 12 D3 4C 00 11 FF
```

Figure 1.4: HEX dump.

1.4.1 UNSIGNED INTEGER CODE

There are different methods for expressing numerical values. The most common in use today is the positional number system. In a base b number system, b symbols are used to represent the first b natural numbers $(0, 1, \ldots b\text{-}1)$. To represent other numbers, the position of the symbol in the number is assigned a weight that is a power of b. An n-digit number $a_{n-1}a_{n-2}\ldots a_1 a_0$ in base-b has the value

$$a_{n-1}a_{n-2}\ldots a_1 a_0 = a_{n-1} \cdot b^{n-1} + a_{n-2} \cdot b^{n-2} \ldots a_1 \cdot b^1 + a_0 \cdot b^0 = \sum_{i=0}^{n-1} a_i \cdot b^i$$

For example, in base 10 the number 712 means 7(100)+1(10)+2(1).

In binary, the base (b) is 2 and the first two natural numbers (0 and (1) are used as **binary digits** or (bits). Using a positional number code, the 8 bit binary number 10100011 has the value

$$10100011_2 = 1(2^7) + 0(2^6) + 1(2^5) + 0(2^4) + 0(2^3) + 0(2^2) + 1(2^1) + 1(2^0)$$
$$= 128_{10} + 32_{10} + 2_{10} + 1_{10} = 163_{10}$$

The weight of each position is either 1 or 0, meaning we add the weight or not to the value of the number. One can simply tally the weights of the non-zero positions to obtain the value. Thus, for example, $10001_2 = 1 + 16 = 17_{10}$.

Computers manipulate numbers with a fixed number of bits (that is with finite precision). Thus, there is a limit to the range of numbers that can be represented. The largest unsigned integer that can be represented by a n-bit binary number can be found by setting all the weights to 1. Since

$$1 \cdot 2^{n-1} + 1 \cdot 2^{n-2} \ldots 1 \cdot 2^1 + 1 \cdot 2^0 = \sum_{i=0}^{n-1} 2^i = 2^n - 1$$

it follows that the range of an n-bit unsigned integer M is

$$0 \leq M \leq 2^n - 1$$

In other words, the largest value of an n-bit unsigned integer is the precision minus 1 (because one codeword is needed to represent 0).

Example 1.8. What is the range of a 6 bit unsigned integer?
Solution: Precision=2^6=64. The maximum value that can be represented is one less or 63.
Answer: range of 6 bit unsigned integer is 0 to 63.

Example 1.9. What is the range of a 12 bit unsigned integer using prefix notation?
Solution: Precision=2^{12}=$2^2 \times 2^{10}$=4Ki. Thus, the maximum value that can be represented is one less or 4Ki - 1.
Answer: The range of 12 bit unsigned integer is 0 to 4Ki-1.

The standard method for converting a number to base b is to repeatedly divide by b; the remainder after each division is taken to be the next least significant digit of the base-b number and the quotient becomes the next dividend. This method is illustrated in Table 1.5 which shows conversion of the decimal number 77 to binary: 77_{10}=1001101_2. The result can be checked by converting back; indeed, 1+4+8+64=77.The most significant bit of a number is sometimes abbreviated msb, while the least significant bit is abbreviated lsb.

Table 1.5: Tabulated values for repeated division base conversion algorithm.

Step	Division	Quotient	Remainder
1	77/2	38	1 (lsb)
2	38/2	19	0
3	19/2	9	1
4	9/2	4	1
5	4/2	2	0
6	2/2	1	0
7	1/2	0	1 (msb)

An alternative algorithm is to repeatedly subtract the weights, starting from the msb, as long as the difference does not become negative. If a weight is subtracted at a given step, the corresponding bit of the number is a 1 (otherwise it is 0). This is illustrated in Table 1.6, which shows the conversion steps to represent decimal 117 as an 8b unsigned binary number.

When converting to binary, it is often faster to first convert to base 16 or base 8 (hexadecimal or octal) and then to binary. This has the advantage of fewer divisions and the number is already in the more efficient hex or octal representation. In addition, divide by 8 can be done quickly without

Table 1.6: Tabulated values for repeated subtraction base conversion algorithm.

Step	Remaining Value to Represent	Weight	Bit	Accumulated Value
1	117	128	0 (msb)	0
2	117	64	1	64
3	53	32	1	96
4	21	16	1	112
5	5	8	0	112
6	5	4	1	116
7	1	2	0	116
8	1	1	1 (lsb)	117

a calculator. Conversion of the number 156 from decimal to binary using octal as an intermediate base is shown in Table 1.7. The resulting octal value is 234_8 which can be quickly converted to the binary value 010011100_2.

Table 1.7: Tabulated values for conversion to octal.

Division	Quotient	Remainder
156/8	19	4 (last significant octal digit)
19/8	2	3
2/8	0	2 (most significant octal digit)

1.4.2   TWO'S COMPLEMENT SIGNED INTEGER CODE

The most common code used to represent signed integers is the two's complement code. In two's complement, the weight of the most significant bit is negative. Thus, the value of an n-bit two's complement number is given as

$$a_{n-1}a_{n-2}\ldots a_1a_0 = -a_{n-1}\cdot 2^{n-1} + a_{n-2}\cdot 2^{n-2}\ldots a_1\cdot 2^1 + a_0\cdot 2^0 = -a_{n-1}\cdot 2^{n-1} + \sum_{i=0}^{n-2}a_i\cdot 2^i$$

The largest negative value is found by setting the msb to 1 and all remaining bits to zero, which results in the value

$$10\ldots 00 = -1\cdot 2^{n-1} + 0\cdot 2^{n-2}\ldots 0\cdot 2^1 + 0\cdot 2^0 = -2^{n-1}$$

which is the negative of half the precision. This result is intuitive because one would expect half of the values to be negative, thus half of the precision is devoted to the negative numbers $-1, -2, \ldots -2^{n-1}$.

The largest positive value represented is found by setting the msb to 0 and all other bits to 1, which gives

$$-0 \cdot 2^{n-1} + 1 \cdot 2^{n-2} \ldots 1 \cdot 2^1 + 1 \cdot 2^0 = \sum_{i=0}^{n-2} 2^i = 2^{n-1} - 1$$

which is half the precision minus one. This follows from analysis of the unsigned integer range, where the minus 1 is due to the need to represent 0.

Example 1.10. What is the range of a 6 bit signed integer?
Solution: Precision=2^6=64. The maximum negative value is the negative of half the precision, -32. The maximum positive value is half the precision minus 1, +31.
Answer: range of 6 bit unsigned integer is -32 to 31.

Example 1.11. What is the range of a 12 bit signed integer using prefix notation?
Solution: Precision=2^{12}=$2^2 \times 2^{10}$=4Ki. Half the precision is 2Ki.
Answer: range of 12 bit signed integer is -2Ki to 2Ki-1.

All negative numbers in two's complement have an msb of 1; all positives have an msb of 0. Thus, the sign of a two's complement number can be determined by simply inspecting its most significant bit (msb).

To negate an n-bit number M, the two's complement operation is used. To compute the two's complement in binary, complement (toggle) all bits and add one to the result (keeping only the least significant n bits after the addition). Thus, if $M = 0111_2$

$$-M = 1000 + 1 = 1001$$

This is also equivalent to the following somewhat simpler procedure: starting at the lsb, copy the bits up to and including the first 1 bit found; then, complement the remaining bits. To encode a signed decimal value in two's complement, it is generally easiest to find the unsigned binary representation of the absolute value of the number and then perform the two's complement operation if the number is negative.

Just as padding 0's to the front of an unsigned number does not change its value, copying the msb of a two's complement signed number does not change its value. This operation, called *sign-extending*, can be easily verified by the following equation

$$a_{n-1}a_{n-1}a_{n-2}\ldots a_1 a_0 = (-a_{n-1}2^n + a_{n-1}2^{n-1}) + \sum_{i=0}^{n-2} a_i \cdot 2^i$$

$$= a_{n-1} \cdot (-2^n + 2^{n-1}) + \sum_{i=0}^{n-2} a_i \cdot 2^i = -a_{n-1} \cdot 2^{n-1} + \sum_{i=0}^{n-2} a_i \cdot 2^i$$

For example, consider the two's complement value 1101. The value is computed as

$$(-8) + 4 + 1 = -3$$

If we extend the sign-bit out one position, we get the 5 bit number 11101, which has the value

$$(-16 + 8) + 4 + 1 = -3$$

Notice the weight of the new sign bit combines with the (now) positive weight of the old sign bit to yield the same value net combined weight (-16+8 = -8). This process can be repeated as many times as desired to extend the size of a two's complement operand without changing its value. Consider, for example, the conversion of 111101:

$$(-32 + 16 + 8) + 4 + 1 = (-16 + 8) + 4 + 1 = (-8) + 4 + 1 = -3$$

Example 1.12.   What is the 8 bit two's complement representation of -30_{10}?
Solution: 30=16+8+4+2, which gives 00011110. The two's complement of 00011110 is found by copying up to the first 1 (starting at lsb) and complementing the remaining bits, which gives 11100010. To verify, summing the weights we get
        -128+64+32+2=-128+98=-30.
Answer: The two's complement representation of -30_{10} is 11100010.

Example 1.13.   Write the two's complement number 10011111 as a 12 bit two's complement number.
Solution: Sign-extend to 12 bits. The sign is 1, so we get 111110011111.
Answer: 100111111 = 111110011111.

Example 1.14.   Write the two's complement number 0101 as an 8 bit two's complement number.
Solution: Sign-extend to 8 bits. The sign is 0, so we get 00000101.
Answer: 0101=00000101.

Example 1.15.   Write the unsigned number 10101 as an 8 bit number.
Solution: Because it is unsigned, we zero-extend to get 00010101.
Answer: 10101=00010101.

1.4.3   ASCII CHARACTER CODE

To represent character data, each letter, punctuation mark, etc. can be assigned a code value. To facilitate interaction among computers of different types, the American Standard Code for Information Exchange (ASCII) has been developed to standardize the character code. ASCII character code is a 7b code that consists of printable and nonprintable control characters, many of which are currently rarely used for their intended purpose. The printable ASCII characters represent characters, numbers, punctuation symbols and a variety of other symbols. Printable ASCII characters are often represented by placing the character in single quotes; for example, 'A' represents the ASCII code for uppercase A (41_{16}). A list of ASCII values is provided in Table 1.8. In the table, the code values are listed in hexadecimal. The printable characters start at 20_{16} and go up to $7E_{16}$. There is an extended ASCII code that uses 8 bits; the first 128 characters are as shown below; the extended characters represent various symbols.

Most CPUs do not provide instructions that manipulate ASCII values directly. However, it should be noted that the code does have some structure that can be exploited by the programmer. For example, the upper case letters, lowercase letters and numbers appear sequentially in the code. This means that code words can be compared. For example, 'A' is less than 'B' and '8' + 1 = '9'. The programmer can exploit these properties of the code when writing a program to manipulate ASCII data.

1.4.4   FIXED-POINT BINARY CODE

Non-integer data can be represented using a similar positional number format; however, this format must be extended to introduce the concept of a binary point. In general, an unsigned binary fixed-point integer can be represented as

Table 1.8: Table of the American Standard Code for Information Exchange (ASCII) code values.

Code	Char.	Code	Char.	Code	Char.	Code	Char.
00	NUL	20	Space	40	@	60	`
01	SOH	21	!	41	A	61	a
02	STX	22	"	42	B	62	b
03	ETX	23	#	43	C	63	c
04	EOT	24	$	44	D	64	d
05	ENQ	25	%	45	E	65	e
06	ACK	26	&	46	F	66	f
07	BEL	27	'	47	G	67	g
08	BS	28	(	48	H	68	h
09	HT	29	)	49	I	69	i
0A	LF	2A	*	4A	J	6A	j
0B	VT	2B	+	4B	K	6B	k
0C	FF	2C	,	4C	L	6C	l
0D	CR	2D		4D	M	6D	m
0E	SO	2E	.	4E	N	6E	n
0F	SI	2F	/	4F	O	6F	o
10	DLE	30	0	50	P	70	p
11	DC1	31	1	51	Q	71	q
12	DC	32	2	52	R	72	r
13	DC3	33	3	53	S	73	s
14	DC4	34	4	54	T	74	t
15	NAK	35	5	55	U	75	u
16	SYN	36	6	56	V	76	v
17	ETB	37	7	57	W	77	w
18	CAN	38	8	58	X	78	x
19	EM	39	9	59	Y	79	y
1A	SUB	3A	:	5A	Z	7A	z
1B	ESC	3B	;	5B	[	7B	{
1C	FS	3C	<	5C	\	7C		
1D	GS	3D	=	5D	]	7D	}
1E	RS	3E	>	5E	^	7E	~
1F	US	3F	?	5F	_	7F	DEL

$$a_{n-1}a_{n-2}\ldots a_1 a_0 \cdot c_{-1}c_{-2}\ldots c_{-m} = \sum_{i=0}^{n-1} a_i \cdot 2^i + \sum_{i=1}^{m} c_{-i} \cdot 2^{-i}$$

And a signed fixed point number as

$$a_{n-1}a_{n-2}\ldots a_1 a_0 \cdot c_{-1}c_{-2}\ldots c_{-m} = -a_{n-1} \cdot 2^{n-1} + \sum_{i=0}^{n-2} a_i \cdot 2^i + \sum_{i=1}^{m} c_{-i} \cdot 2^{-i}$$

In other words, the bits to the right of the *binary point* are assigned weights that are negative powers of 2 (1/2, 1/4, 1/8, etc.). This is actually a generalization of the positional number codes described above. When using this format the computer has no concept of the binary point; thus the binary point is implied by the program that manipulates the data. It is important to remember that the binary point is not stored with the data.

As an example, consider the unsigned fixed-point 8b value 10101100. Assume that there are 4 *binary places* (i.e., the binary point occurs between bits 3 and 4). The value represented by this number is

Bit	1	0	1	0	.	1	1	0	0	
×Weight	8	4	2	1		$\frac{1}{2}$	$\frac{1}{4}$	$\frac{1}{8}$	$\frac{1}{16}$	
	+8		+2			$+\frac{1}{2}$	$+\frac{1}{4}$			= 10.75

Assume now that the same unsigned fixed-point 8b value 10101100 has only 3 binary places. The value is computed as

Bit	1	0	1	0	1	.	1	0	0	
×Weight	16	8	4	2	1		$\frac{1}{2}$	$\frac{1}{4}$	$\frac{1}{8}$	
	+16		+4		+1		$+\frac{1}{2}$			= 21.5

The implied position of the binary point is crucial to determining the value of the data. The astute observer might realize that the value 21.5 is twice the value 10.75. This is not by accident; just as moving the decimal point to the right in base 10 is equivalent to multiplication by 10, so moving the binary point to the right in base 2 is equivalent to multiplication by two. In the same manner, moving the binary point to the left by 1 is equivalent to division by 2. If one moves the binary point all the way to the right of the last bit, what remains is an integer whose value is the fixed point value multiplied by 2^m, where m is the number of binary places moved. Thus, the value of a fixed point number can be found by taking the integer value of the data and dividing it by 2^m. This property can be exploited even further: the bits to the left of the binary point can be interpreted as an integer value; then, the bits to the right of the binary point can be interpreted as an integer and the value found by dividing by 2^m. For example, the unsigned fixed point number 101.011 has the integer 5 to the left of the binary point and the integer 3 to the right. Since there are 3 binary places, the value of the number is 5 $3/2^3$ = 5 3/8. This is perhaps the fastest method to manually interpret the value of a fixed-point number.

Example 1.16.   What is the value of the unsigned, fixed-point binary number 01101001, if the number of binary places is taken to be 3?
Solution:
    Method 1: The integer value of the operand is 1+8+32+64=105. With 3 binary places, we divide this by 2^3=8 to get the fixed point value of 13.125.
    Method 2: 01101.001 has integer value 13 to left of binary point and 1 to the right; thus, the number is 13 $1/2^3$ = 13 1/8.
Answer: 01101.001 = 13.125.

Example 1.17.   Represent the value 3 1/8 using 8-bit fixed-point binary representation with 5 binary places.
Solution: 5 binary places means the fractional part is in 32^{nds}. Thus, we represent 3 1/8 as 3 4/32. There are 8 - 5 = 3 bits for the integer part (011). The fraction part has 5 bits, so it is 4/32 (or the integer 4 expressed using 5 bits). Thus, we get 011.00100.
Answer: 3 1/8 is expressed as 011.00100.

Example 1.18.   What is the value of the signed, fixed-point binary number 10101001, if the number of binary places is taken to be 4?
Solution:
    Method 1 (sum the weights): 1010.1001 = -8 + 2 + 1/2 + 1/16 = -5 7/16.
    Method 2 (two's complement): two's complement of 1010.1001 is 0101.0111 (ignore the binary point and perform the two's complement operation as before). Now the number can be read by inspection to be 5 7/16, so the answer is -(5 7/16).
    Method 3: we can read the number to the left of the binary point as two's complement integer, and add the unsigned fraction to the right. Thus, 1010 is the integer -6. To the right of the binary point, we have 1001, which is 9. Therefore the number has the value -6 + 9/16 = -(5 7/16).
Answer: 1010.1001 = -5 7/16.

Example 1.19. Express the value of the 10b unsigned fixed-point binary number 1000101001 as a proper fraction and an improper fraction, assuming there are 7 binary places.
Solution: The integer value of the data is 1+8+32+512=553. Thus, as an improper fraction, the value is $553/2^7$ (or 553/128).

As a proper fraction, the integers to the left and right of the binary point are evaluated. This, 100.0101001 has the value of 4 to the left of the binary point and 41 to the right. Thus, the value is $4\ 41/2^7$ (or 4 41/128).
Answer: 100.0101001 = 553/128 = 4 41/128.

Example 1.20. Represent $1\frac{1}{3}$ using 8-bit fixed-point binary representation with 4 binary places.
Solution: 4 binary places means the fractional part is in 16^{ths}; setting $\frac{1}{3} = \frac{N}{16}$ we get $N = 5.33$. Because N is not an integer, there is not enough precision to completely express $\frac{1}{3}$. We truncate N to an integer; thus, $N = 5$. 4 bits are used for the integer part (0001).
Answer: $1\frac{1}{3}$ is expressed as 0001.0101.

1.4.5 BINARY CODED DECIMAL

Binary coded decimal (BCD) is a 4b code used to represent base 10 numbers. The valid BCD codes are 0000 for 0; 0001 for 1; 0010 for 2; 0011 for 3; 0100 for 4; 0101 for 5; 0110 for 6; 0111 for 7; 1000 for 8; and 1001 for 9. The remaining code words, 1010–1111, represent invalid values. BCD is a code that allows fast access to the decimal digits of a value (in binary form).

Each digit of a BCD number is encoded separately. Most computers store data as bytes in memory; thus it is possible to fit two BCD digits into a single byte. When two digits are packed into a single byte the code is referred to as packed-BCD. Unpacked BCD (or simply BCD) stores each BCD-encoded digit in its own byte.

Arithmetic operations on BCD values are often not directly supported by the CPU; thus, the programmer must apply algorithms to perform the operations.

1.5 ADDITION AND SUBTRACTION OF POSITIONAL NUMBER CODES

Addition and subtraction of unsigned numbers using the positional number code follows the familiar "pencil-and-paper" algorithm we use for manual addition of decimal numbers. Two's complement signed numbers use the same algorithm. Two important points to remember are:

Point #1 A computer always performs the arithmetic using a fixed number of bits – such as 8b addition or 16b subtraction. If the result exceeds the range of the code being used to represent the numbers, overflow can occur. In addition, if the size of the operands differs, the smaller operand must be sign-extended or zero-extended as appropriate before performing the arithmetic.

Point #2 A computer performs addition and subtraction of binary values; it does not take into consideration the binary point or the code being used. Thus, the programmer must be careful when operating on signed and unsigned or on integer and fixed-point numbers in the same operation.

Consider the binary addition in Figure 1.5, which is performed using the standard addition algorithm. The addition is an 8b addition. Also shown are how the addends and sum are interpreted

				Unsigned	Two's Complement	Fixed-Point (3 binary Places)
Carry Bits	1 1 1 1 1					
	0 0 1 0 1 0 1 1			43	43	5.375
+	1 0 1 1 1 0 0 1			+185	+ -71	+23.125
Sum	1 1 1 0 0 1 0 0			228	-28	28.500

Figure 1.5: Illustration of addition of binary numbers with no overflow.

using 3 different codes: unsigned integer, signed integer and unsigned fixed-point binary with three binary places. The carry bits in each column represent surplus from the addition of the column to the right (the next least significant column).

Despite that the values in Figure 1.5 are interpreted using 3 different codes, the results are all correct. In each of the three cases, the value of the result of the operation is within the range of the 8b code being used. Now consider the binary addition in Figure 1.6. In this case, there is a

				Unsigned	Two's Complement	Fixed-Point (3 binary Places)
Carry Bits	1 1 1 1 1 1 1					
	1 0 1 0 1 0 1 1			171	-85	21.375
+	1 1 1 1 1 0 0 1			+249	+ -7	+31.125
Sum	1 0 1 0 0 1 0 0			X 164	-92	X 20.500

Figure 1.6: Illustration of addition of binary numbers with overflow.

carry out of the most significant bit position, but since it is an 8 bit addition and this carry would produce a 9th sum bit, it is not included in the sum (recall point #1 above). In this instance, the binary addition does not produce the correct result when the operands are interpreted as unsigned integer or unsigned fixed-point values. This is because for these two codes, the actual value of the result of the operation is too large to fit in the range of the 8b code being used. For example, 171+249 is 420; the range of an 8b unsigned integer, however, is 0 to 255. Thus, the result of the unsigned addition is too large to be represented with an 8b code. A carry out of the msb indicates overflow when performing unsigned addition.

When the binary data are interpreted as two's complement values, however, the result of the operation is correct. Clearly, a carry out of the msb does not indicate overflow for two's complement addition. For two's complement addition, one can manually determine overflow by looking at the signs of the operands and the sign of the result. If adding two positive numbers yields a negative result, then there is overflow. If adding two negative numbers yields a positive result, then there is overflow. When adding a positive to a negative, overflow can never occur. These three conditions can be shown to be logically equivalent to the following: if the carry into the msb does not equal the carry out of the msb, then there is overflow. This is how the CPU determines when two's complement overflow occurs. Note that in the above example, there is a carry bit into the msb and a carry out, indicating no overflow.

Overflow for subtraction can be determined in a similar way. For subtraction, a borrow bit in column i indicates a deficit in column $i-1$. Thus, the subtraction $0-1$ is said to produce a borrow out while the subtraction $1-1$ does not. For unsigned subtraction, if there is a borrow out of the msb then overflow occurs. For signed numbers, the test for subtraction overflow is similar to the test for unsigned numbers. Subtracting a negative value from a positive value should never yield a negative result. Similarly, subtracting a positive value from a negative value should never yield a positive result. An example subtraction is shown in Figure 1.7. Note that in the figure, each unsigned subtraction

									Unsigned	Two's Complement	Fixed-Point (3 binary Places)
Borrow Bits	1	1	1	1							
	0	0	1	0	1	0	1	1	43	43	5.375
-	1	0	1	1	1	0	0	1	-185	- -71	-23.125
Difference	0	1	1	1	0	0	1	0	114	+114	14.250

Figure 1.7: Example of binary subtraction with no overflow.

results in overflow (there was a borrow out of the msb). The signed subtraction had a borrow into the msb and a borrow out of the msb, indicating no overflow. Table 1.9 summarizes the overflow conditions for binary addition and subtraction.

Binary addition and subtraction can also be performed directly in octal or hexadecimal base. The binary value of the result and conditions for overflow are the same.

1.6   CHAPTER PROBLEMS

1. Define the terms computer, algorithm and program, general-purpose computer, embedded computer, CPU and I/O.

2. True or false: A microcontroller is a microcomputer.

Table 1.9: Summary of conditions for determining overflow for addition or subtraction of signed or unsigned numbers.

	Addition	*Subtraction*
Signed Operands	Overflow if carry into msb does not equal carry out of msb.	Overflow if borrow from msb does not equal borrow into msb.
	-or-	-or-
	Can also be determined by evaluating the signs of the operands and sign of the result.	Can also be determined by evaluating the signs of the operands and sign of the result.
Unsigned Operands	Overflow if carry out of msb.	Overflow if borrow out of msb.

3. Where does the term bit come from?

4. What are the 3 main components of a computer? What interconnects them?

5. What do RISC and CISC stand for? Describe the difference between them.

6. What is the difference between machine code and assembly language? How is assembly language converted to machine code?

7. What are the basic steps a CPU performs during instruction sequencing?

8. What is the difference between an interrupt and an exception?

9. A CPU uses 10 bit memory addresses. What is the size of its address space?

10. Describe the difference between volatile and non-volatile memory. List one type of each.

11. What is the main functional difference between a ROM and EEPROM? Which type would be more suited for storing the odometer reading in an automobile?

12. What is the precision of a 5b code? 10 bit code? 11b code? (write the result as a number).

13. What size binary code would be needed to store

    a. The minutes of the current time?

    b. The number of miles between your current position and any point on earth (the circumference of the earth is approximately 24,902 miles)

14. Exactly how many bytes are in 3 Mibytes? 1 GiBytes? (write the result as a number).

15. Use prefix notation to describe 2^{25} Bytes.

16. What is the precision of a 32 bit code in prefix notation?

17. How much increase precision do you get when you add one more bit to a code (for example, when going from using an n bit code to an $n+1$ bit code)?

18. Complete the following table; in each case, use the smallest number of digits/bits possible.

Binary	Octal	Decimal	Hexadecimal
10011111	237	159	9F
	073		
		1772	
			FEED
0101010011			

19. What is the value represented by the binary data 01100110 assuming

    a. ASCII

    b. Packed BCD

    c. Unsigned integer code

    d. Signed integer code

    e. Unsigned fixed-point (two binary places)

    f. Signed-fixed point (6 binary places)

20. What is the range of a 11b code, using

    a. Unsigned binary

    b. Signed binary

    c. Unsigned Fixed-point with 8 binary places

    d. Signed fixed-point with 8 binary places

21. Convert to unsigned binary using the fewest number of bits possible.

    a. 127

    b. 128

    c. 33

22. Convert to signed binary using the fewest number of bits possible.

    a. +13

    b. -13

    c. +32

    d. -32

    e. -7

23. Represent each of the following using an 8b unsigned fixed-point notation with 4 binary places; express the result in binary and hex notation.

    a. $3\frac{1}{4}$

    b. 3.75

    c. 117/16

    d. 1/5

    e. Pi

24. Repeat Problem 23, representing the negative value of each number using an 8b signed fixed-point notation with 4 binary places.

25. Add each of the following and determine if there is signed overflow and unsigned overflow.

    a. 11110000+00100000

    b. 01110000+01011111

    c. 11110110+11111100

    d. 8F2+72E

26. Subtract each of the following and determine if there is signed overflow and unsigned overflow.

    a. 0101−1110

    b. 10000−01111

    c. 74F−6FE

27. Add the signed values

    a. 10111+011

    b. 111110+1111

    c. EF+7A0

    d. 7E+FFF

28. In ASCII code, write the values to represent

    a. EECE252

    b. 742

    c. Computer

29. Given the HEX dump

```
0080   00   01   34   44   99   AF   FE   FF
0088   00   33   29   AA   BC   CC   D0   08
```

    a. What is the value of the byte at address $0087?

    b. What is the value of the byte at address $008B?

    c. What is the address of the byte whose value is 44?

    d. What is the address of the byte whose value is 08?

30. Sketch a memory map of the hex dump shown in Problem 29.

CHAPTER 2

Programmer's Model of the HCS08 CPU

The goal of this section is to become familiar with the instruction set architecture of the HCS08 CPU family. The focus is on introducing the building blocks needed to write assembly language programs but not on the programming itself, which will be covered in the next chapter. All of the instructions in the HCS08 CPU instruction set are listed; however, attempting to cover all details would be excessive. Where more information is needed, refer to the Freescale *HCS08 Family Reference Manual*.

2.1 INTRODUCTION

The HCS08 CPU is the central processor unit of the Freescale Semiconductor (formerly Motorola) M68HSC08 family of microcontrollers. The HCS08 CPU is very similar to the instruction set of its predecessor, the HCS08 CPU, allowing software written for the older CPU to run without modification on the new CPU while providing increased performance and additional features. The HCS08 CPU, in turn, was designed to be backwards compatible with the older M68HC05. Such upgrade paths are common within CPU families; as technology improves, CPUs must keep pace to remain competitive. Backwards compatibility eases the transition to a new CPU without losing the investment in the existing software base.

  The HCS08 CPU is an 8 bit architecture, which means that it primarily includes support to manipulate 8 bit operands. The HCS08 CPU does have limited support to manipulate data of sizes other than 8 bits. The HCS08 CPU uses 16 bit addresses to access memory, indicating that it can address up to 64 kibibytes of memory. Registers used to hold memory addresses are therefore 16 bits wide, while those intended primarily for data are 8 bits.

  The HCS08 CPU is an accumulator-based architecture. The results of arithmetic and logical operations are accumulated in CPU registers. For example, an addition operation on an accumulator-based architecture would have the form

$$A = A + \text{operand}$$

where A is a CPU register and the operand is typically data in memory. The principle advantage of this form of operation is that each instruction only needs to specify two operands: the destination/left source operand and the right source operand. In general-purpose register architectures, three operands must be specified (destination, left source and right source). Accumulator-based architectures therefore tend to have more compact machine code, which is one of the reasons embedded microcontroller architectures, like the HCS08 CPU, tend to be accumulator-based.

2.2   REGISTERS

The HCS08 CPU registers are shown in Figure 2.1. Register A is an 8 bit accumulator register. This general purpose register can be used to hold operands and accumulate the results of arithmetic and logical operations. The bits of accumulator A are numbered left-to-right from 7 down to 0, with bit 7 representing the most significant bit (msb) and bit 0 the least significant bit (lsb).

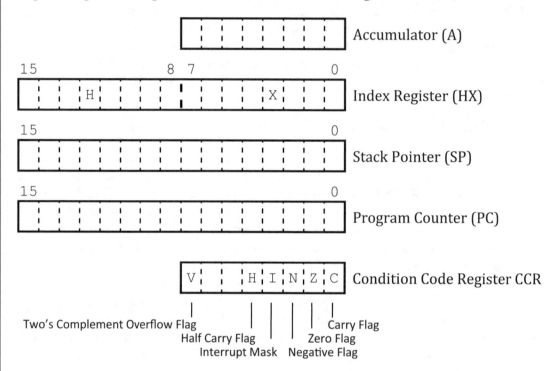

Figure 2.1: HCS08 CPU Registers and flags.

To demonstrate the accumulative nature of the register, the assembly language instructions to compute $45_{10} + A0_{16} + 01100010_2$ are:

```
LDA   #45           ; Load A instruction: A=45
ADD   #$A0          ; Add to A instruction: A = A + $A0
ADD   #%01100010    ; A = A + %01100010
```

The first instruction, LDA #45, places the constant 45_{10} into accumulator A. LDA is the mnemonic for load into accumulator A. The '#' prefix indicates that the operand is a constant.

Everything that appears after the ';' on a line is a comment (the ';' and comment are optional). The second and third instructions are ADD, which is the mnemonic for "add to A." The second instruction adds to the content of A the constant $A0_{16}$; '\$' is the prefix indicating that the constant is written in hexadecimal. In the third instruction, the '%' is the prefix to specify the base of the constant as binary. Note that the assembly language is not understood by the CPU; eventually, it must be converted to machine code by an assembler. In the machine code, all three constants will be encoded in binary form. All arithmetic and logical instructions on A share the accumulative property: the left-hand operand and result of the operation are understood to be in accumulator A. Because A is used to accumulate results of arithmetic and logical instructions, the name "accumulator A" is used to describe this register, rather than register A.

The index register HX is a 16 bit (2 byte) register. An index register is used to index or "point to" an operand in memory. This is accomplished by placing the address of the memory operand in HX. The upper and lower bytes of HX can also be accessed independently as H or X. This is because HX is actually composed of the two registers H and X, a result of the upgrade from the CPU05 architecture, which had only an 8-bit index register called X. When not being used to point to memory operands, HX can be used to temporarily hold operands; in addition, HX or X can be used as simple 16b or 8b counter registers, though not at the same time.

The stack pointer SP is a special index register that points to the top of the stack. The stack is a region of RAM that is used as a last-in-first-out (or LIFO) queue. Most CPUs contain a stack pointer to facilitate development of subroutines and to support interrupts. Use of the stack will be covered in more detail in the next chapter. For now, it is important to remember that this register is dedicated for this purpose and generally cannot be used for other purposes.

The program counter is another special purpose index register that points to the next instruction to be executed in memory. After each instruction fetch, the PC is incremented by the number of bytes in the machine code of the instruction, making it point to the next sequential instruction in memory. In order to implement conditional execution, branch and jump instructions can be used to make the PC point to other instructions. Because the PC has dedicated use in instruction sequencing, it, like the stack pointer, cannot be used to hold other types of data.

The final register in the HCS08 CPU instruction set is the Condition Code Register (CCR), also known as the flags register. This register contains 5 one bit flags (V, H, N, Z, C) and the interrupt mask bit (I). The flags are used to indicate information on the outcome of arithmetic and logical operations, while mask bits are used to enable or disable events or actions. The carry flag C is the general purpose flag of the CPU; it is used in a variety of ways by different instruction types. For arithmetic and logical operations, C is generally used to hold and extra bit of the result when it is too large to fit in the space allowed for its destination. For example, C is used to hold the carry or borrow out of addition or subtraction operation. For shift operations, it holds the bit shifted out of the operand. V is the two's complement overflow flag, which the CPU sets when arithmetic operation results in two's complement overflow. The half-carry flag (H) is set when an addition has a carry from lower nibble to upper nibble or a subtraction results in a borrow out of the lower nibble; this flag is used primarily for binary-coded decimal (BCD) operations. The negative flag (N) is set when an operation produces a negative result (i.e., N reflects the msb of the result). The zero flag (Z) is set if an operation results in zero. The I mask bit is used to control interrupts. When I is set, interrupt requests are masked (blocked).

Example 2.1. Accumulator A contains 34_{16}. The instruction ADD #$70 is executed by the CPU. What are the values in accumulator A and the flag registers after the instruction executes?
Solution: The instruction performs the 8b addition of 34_{16} to 70_{16}, which results in $A4_{16}$

$$
\begin{array}{ccccccccc}
 & 1 & 1 & 1 & & & & & \\
 & 0 & 0 & 1 & 1 & 0 & 1 & 0 & 0 \\
+ & 0 & 1 & 1 & 1 & 0 & 0 & 0 & 0 \\
\hline
 & 1 & 0 & 1 & 0 & 0 & 1 & 0 & 0 \\
\end{array}
$$

Looking up the ADD instruction in the *HCS08 Family Reference Manual*, it is found that the ADD instruction affects all 5 CPU flags. Because there is no carry out of the msb, C is cleared (C=0). The carry into the msb does not equal the carry out; therefore there is two's complement overflow; V is set (V=1). There was no carry out of bit 3, so H is cleared (H=0). The msb of the result is 1; therefore, N is set (N=1). Finally, the result is not zero, so Z is cleared (Z=0).
Answer: A contains $A4_{16}$. C=0, V=1, H=0, N=1, Z=0.

2.3   MEMORY

Each member of the HCS08 family of microcontrollers has a specific amount of RAM and flash memory as well as a particular set of peripherals and associated I/O interfaces. As a result, there are differences in the locations of memory and memory-mapped I/O ports among the various processors. To minimize the impact of these differences, the HCS08 programmer's model defines five memory regions corresponding to I/O registers (2 separate regions), RAM, flash memory and vectors. These regions are shown in the memory map in Figure 2.2. The region that extends from address $0000 to address $00XX (where XX is 1F, 3F, 5F or 7F depending on the specific family member) are memory-mapped registers corresponding to I/O port interfaces. Software can access the values in these registers in order to communicate with and control I/O devices. Immediately following this region and extending up to at most $017F is static RAM used for programs and data. After RAM, beginning at $0180, there is another block of I/O interface registers. The remainder of the memory map, from $1C00 to $FFFF, is used to map various types of non-volatile memory. The last 64 bytes of this region is further reserved for the interrupt vector table, which holds pointers to the start of the service routines for individual interrupts and exceptions.

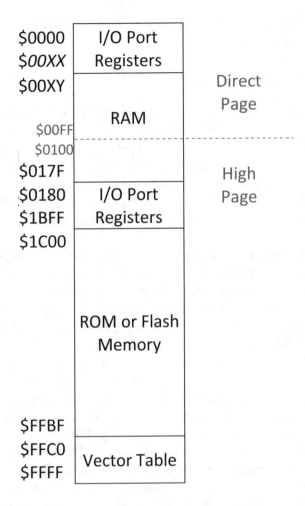

Figure 2.2: HCS08 CPU Family Common Memory Map.

Note that there is a logical boundary that divides the address space, between $00FF and $0100. The 256 address range from $0000-$00FF is called the direct page of the memory (this is also commonly known as a zero page in other types of processors). Since all addresses in the direct page begin with the byte $00, it is possible in some instances to use only one byte of machine code to specify an address in this range. This makes access to memory in the direct page slightly faster than access to the remainder of the address space (which is sometimes referred to as the high page). Depending on the amount of RAM a specific HCS08 CPU family member has, some RAM will be in the direct page and some in the high page. Similarly, some I/O registers will be in the direct page and others in the high page. Access to direct page RAM and I/O interface registers is faster, so its use is preferred when possible.

2.4   ADDRESSING MODES

The addressing modes of a CPU define the methods by which CPU finds the operands of an instruction. In general, operands can be found in memory, in registers or can be constants stored in the machine code of the instruction. For memory operands, the address of the operand is called the *effective address*, which is computed by the CPU based on the addressing mode specified in the machine code. A typical instruction must specify the location of the source operands on which the operation is to be performed, as well as a destination to which the result should be stored. For accumulator-based instructions, the destination and left source operand reside at the same location in register or memory.

The HCS08 CPU has 16 addressing modes; however, not every instruction allows every addressing mode. To determine the allowed modes for each instruction, consult the *HCS08 Family Reference Manual*. In assembly language, the addressing mode is chosen by using a specific syntax when specifying the operands. In machine code, the addressing mode is encoded in the opcode.

Sometimes, operands stored in memory are larger than one byte. When this occurs, the operand is stored in consecutive memory bytes and the effective address of the operand is the address of the first byte (lowest address) in the sequence. There is still a choice, however, as to what order the bytes can be stored in. Consider a 16b operand in memory whose effective address is $0080 and whose value is $ABCD. Figure 2.3 shows the two ways that multi-byte data can be stored: most significant byte (MSB) first or least significant byte (LSB) first (notice that when referring to bytes, capital letters are used; thus while msb is most-significant bit, MSB is most-significant byte). When the MSB of the operand is stored at the effective address, the CPU is said to use big-endian byte ordering, since the big end of the data is what the CPU finds at the address; otherwise, the CPU is using little-endian byte ordering. There are advantages to each method and there is no preferred method – different CPUs use different byte ordering, which can sometimes cause issues with data exchanged between different systems. Freescale microcontrollers use big-endian byte ordering. Intel processors, for example, use little-endian.

Example 2.2.   The 8 consecutive bytes in memory, starting at address $0080, are $FF, $AB, $01, $34, $45, $00, $87, and $FE. The HCS08 CPU loads the 16b operand from effective address $0083 into index register HX. What is the value in HX after the load?
Solution: The byte at location $0083 is $34 and the byte at the next location ($0084) is $45. Since the HCS08 CPU uses big-endian byte ordering, HX contains the 16b value $3445.
Answer: HX contains $3445.

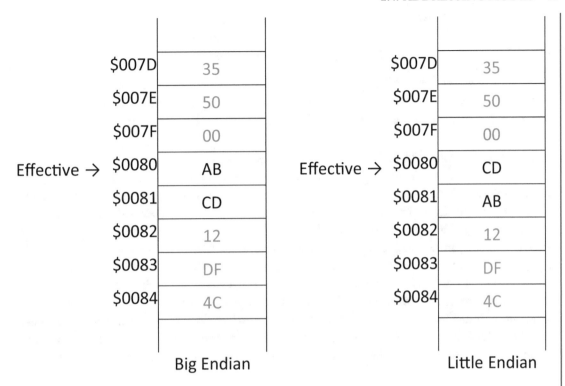

	Big Endian			Little Endian
$007D	35		$007D	35
$007E	50		$007E	50
$007F	00		$007F	00
Effective → $0080	AB		Effective → $0080	CD
$0081	CD		$0081	AB
$0082	12		$0082	12
$0083	DF		$0083	DF
$0084	4C		$0084	4C

Figure 2.3: Little versus Big Endian byte ordering.

2.4.1 INHERENT ADDRESSING

Inherent addressing, the simplest addressing mode, refers to operands that are not explicitly specified but are implied by the instruction. An example is the CLRA instruction, which clears (sets to zero) accumulator A. No operands need to be explicitly listed when using this instruction; both the operand 0 and the destination A are implied by the opcode.

2.4.2 IMMEDIATE ADDRESSING

With immediate addressing, the operand is a constant contained in the machine code of the instruction. Thus, the CPU does not need to locate or fetch the operand from memory – it is immediately available inside the CPU after the instruction is fetched. In HCS08 CPU assembly language, immediate operands are specified by prefixing them with a '#' symbol followed by one of the base prefixes '$' (hexadecimal), '%' (binary), '@' octal or no prefix for decimal. The following instruction sequence (which is copied from above) illustrates the use of immediate addressing. This sequence computes $45_{10} + A0_{16} + 01100010_2$.

```
LDA    #45           ; Load A instruction: A=45
ADD    #$A0          ; Add to A instruction: A = A + $A0
ADD    #%01100010    ; A = A + %01100010
```

As a practical point, this sequence has been provided to demonstrate the use of immediate addressing. It could be argued that the above sequence actually computes nothing, since it only manipulates constants; the same thing could be achieved using the single instruction LDA #$47.

2.4.3   DIRECT AND EXTENDED ADDRESSING

Direct and extended direct (or simply extended) addressing are forms of what is sometimes referred to as absolute addressing: the memory address of the operand is contained in the machine code of the instruction. In direct addressing, only 8 bits of the address are specified; the upper 8 bits of the address are implied as $00. Extended addressing uses full 16 bit addresses. Because the address is stored in the machine code, direct addressing saves a byte of machine code each time it is used, resulting in smaller programs (which is important in embedded systems with small amounts of memory). Of course, direct addressing is limited to operands in the direct page (memory in the address range $0000–$00FF), while extended addressing can refer to an operand anywhere in memory ($0000–$FFFF). Thus, direct addressing provides fast access to the zero page, while extended addressing must be used for operands outside of the zero page. Figure 2.4 illustrates the operation of extended direct addressing. In the figure, the instruction register represents an internal

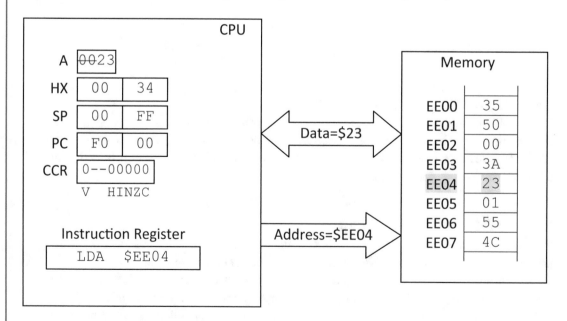

Figure 2.4: Illustration of extended and direct addressing.

register that the CPU uses to hold the machine code of the current instruction; this register is part of the implementation of the processor and not part of the programmer's mode (i.e., the programmer cannot access this register). The registers are shown with their content in hexadecimal; to indicate a change in the content of the register, the old value is shown with a line through it and the new value shown to the right. Thus, in Figure 2.4 the content of accumulator A before execution is $00; after execution, it is $23. The address and data busses are shown with the values that would appear on them during execution of the instruction. In this example, the effective address is $EE04; this address appears on the address bus, selecting the appropriate byte from the memory (shown highlighted). Since the operation being performed is a load, the value at this memory location ($23) appears on the data bus and, because the instruction is LDA, the CPU places this value in accumulator A.

The assembler syntax for direct and extended addressing is similar to that for immediate addressing, but the prefix '#' is omitted. A common mistake when specifying immediate operands is to forget the '#', in which case the assembler interprets the operand as using direct or extended mode. An example of using direct and extended mode is shown below. This instruction sequence loads an 8b value from memory location $EEFE, adds three to it and stores the 8b result to the byte at $0086.

```
LDA   $EEFE   ;copy the byte at memory address $EEFE into A (extended)
ADD   #$03    ;add 3 (immediate)
STA   $86     ;store the result to memory at address $0086 (direct)
```

TIP: A common error is to omit the '#' when using immediate operands. When '#' is omitted, the assembler erroneously interprets the operand as a direct or extended address instead of an immediate constant.

Example 2.3. The 8 consecutive bytes in memory, starting at address $F080, are $FF, $AB, $01, $34, $45, $00, $87, and $FE. What are the addressing mode, effective address, size, and value of the operand in each of the following instructions (Hint: LDA loads its operand into accumulator A and LDHX loads its operand into index register HX).

a)   LDA $F083

b)   LDHX $F083

Solution:
a) For LDA, the addressing mode is extended because the address of the operand ($F083) is greater than $00FF. The effective address is $F083. Since the value is being loaded into accumulator A, the size of the operand is 1 byte. The value of the byte at $F083 is $34.
b) LDHX uses the same addressing mode, so again the mode is extended and the effective address is $F083. Because the operand is to be loaded into HX, it must now be 2 bytes (16 bits). In big-endian byte order, the value of the 16b word at $F083 is $3445.

Answer: a) Extended, effective address is $F083, size is 1B, operand value is $34
b) Extended, effective address is $F083, size is 2B, operand value is $3445

2.4.4   INDEXED ADDRESSING

With indexed addressing, HX contains the effective address of the operand in memory. The effective address is contained in index register HX and must have been explicitly placed there by some other instruction(s). Another way to view this addressing mode is that HX points directly to the operand in memory. This addressing mode is useful when a memory operand needs to be referred to repeatedly, because it can save 2 bytes of machine code (compared to using extended addressing) each time it is used. It also allows a program to manipulate and dereference pointers. A pointer is a data type, found in many high-level programming languages, whose value "points to" an operand in memory by referring to its address. Obtaining the value of the data being pointed to by a pointer is called *dereferencing* the pointer; when indexed addressing is used, the pointer value contained in HX is dereferenced to obtain the value of the operand in memory.

To use indexed addressing in assembly language, the notation ", X" is used for the operand; the comma indicates to the assembler that the mode is indexed (to distinguish it from X, which could be a label). Figure 2.5 illustrates how indexed addressing uses HX to point to the operand in memory. In the figure, the CPU has just completed execution of the instruction "LDA ,X", which loads an

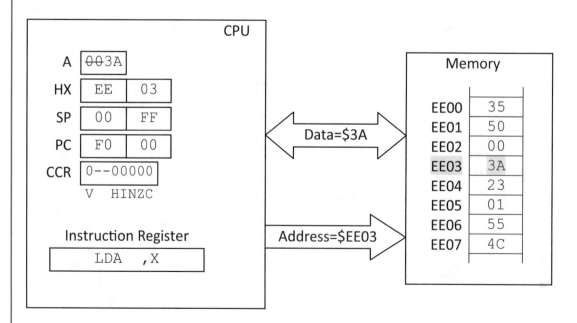

Figure 2.5: Illustration of indexed addressing.

operand from memory into accumulator A using indexed addressing. HX contains the value $EE03, which the CPU uses as the effective address of the operand. The byte value in memory at $EE03 is $2A; this value is loaded into accumulator A.

An example of using indexed addressing is shown below. This example multiplies the value of the byte in memory at address $5211 by 3 and stores the result to memory location $0082; each addressing mode discussed so far is exercised. Note that the multiplication is accomplished by accumulating the value three times. It should be noted that the example below is given to demonstrate the use of several addressing modes and is not the most effective way to perform the computation indicated. A more effective approach is shown in Section 2.7.

```
LDHX    #$5211   ;HX=$5211 (immediate) HX contains address
CLRA             ;clear a (inherent)
ADD     ,X       ;add to A the byte in memory at
ADD     ,X       ;address in HX (indexed)
ADD     ,X       ;do this three times
STA     $82      ;store result to location $0082(direct)
```

2.4.5   INDEXED WITH POST-INCREMENT ADDRESSING

Indexed with post-increment uses exactly the same effective address as indexed addressing – the effective address is contained in HX. After dereferencing the operand, however, the CPU increments the value in HX. Each time post-increment mode is used, the pointer in the index register is advanced to point one byte ahead in memory. This mode is illustrated in Figure 2.6, which shows the contents of the CPU registers after the LDA instruction executes (where important for clarity, the contents of registers before execution are shown with lines through them). The assembler syntax for the mode is ,X+ (note the leading comma).

This mode is generally used for stepping through byte arrays and tables. The HCS08 CPU has only two instructions that allow indexed with post-increment addressing (CBEQ and MOV). These two instructions suggest that the mode is used for searching and copying byte arrays in memory. Note that the HCS08 CPU does not allow this mode with the LDA instruction, but the LDA instruction was used in Figure 2.6 for clarity of presentation.

2.4.6   OFFSET INDEXED ADDRESSING

Offset indexed addressing is similar to indexed addressing in that the index register HX is used as a pointer into memory. With offset indexed mode, however, HX does not point directly to the operand; an *unsigned* constant is used to offset the memory location pointed to by HX. Thus, the effective address is the sum of HX and the offset. This constant is contained in the machine code and can be 8 bit or 16 bits – thus there are two offset indexed modes: 8b – and 16b – offset indexed. This mode can be used to implement data structures, where HX is set to point to the base (start) of the data structure and the offset selects an element from the data structure. Hence, HX is called the *base register*.

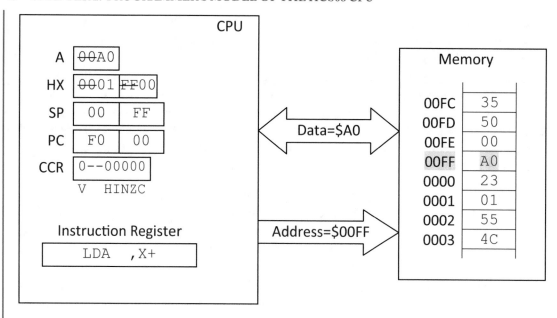

Figure 2.6: Illustrated operation of indexed with post-increment addressing.

Although the offset can be 8 or 16 bits, most assemblers will determine the smallest offset to use in the machine code; the assembly language programmer does not have to specify the size of the offset. When the machine code contains an 8 bit offset, the offset is zero-extended before it is added to HX to generate the effective address of the operand (i.e., the offset is treated as an unsigned offset).

Figure 2.7 illustrates the use of 8 bit offset indexed addressing. HX contains $EE03; the operand $02,X means the effective address (EA) of the operand is computed as EA=HX+$0002, which is $EE03+$0002=$EE05. The operand value is $01, which is loaded into accumulator A.

Example 2.4.   HX points to a table of 8b unsigned integers in memory. Write an instruction to load the 5th element from the table into accumulator A.
Solution: Since HX points to the first element of the table, its offset would be zero. Thus, the fifth element would have an offset of 4. Therefore, the instruction to load the 5th element is

LDA 4,X

Answer: LDA 4,X

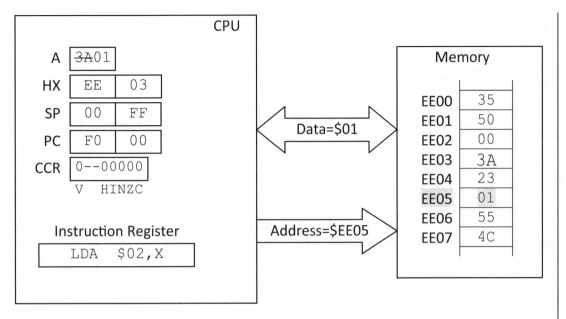

Figure 2.7: Operation of 8-bit offset indexed addressing.

Figure 2.8 illustrates the use of a 16b offset to effectively achieve a negative offset. Recall that the offset is considered as an unsigned offset. In this example, the 16b offset $FFFE (which is the 16 bit two's complement representation of -2) is used as an offset to HX. Because the addition of signed and unsigned follows the same algorithm and due to the limited (16b) precision of the effective address, the result achieved is equivalent to using a negative offset. The effective address is computed as HX+$FFFE, which is EA=$EE03+$FFFE=$EE01.

Some assemblers will accept the notation −2,X to achieve a negative offset; the programmer must remember, however, that even a small negative offset requires 16 bits to encode. For example, the addressing mode $FE,X could be interpreted as either an offset of $00FE or $FFFE, depending on the default behavior of the assembler, leading to ambiguity. When in doubt, specify the full 16b offset.

Example 2.5. Suppose that the time of day, using 24-hour time in the format *hhmmss*, is stored in three consecutive memory bytes (at addresses $F000, $F001, and $F002) using packed BCD. For example, 01:42:36pm would be stored as the bytes 13_{16}, 42_{16}, 36_{16}. Index register HX points to the time data structure in memory (i.e., HX contains $F000). Write an assembly language instruction to load the

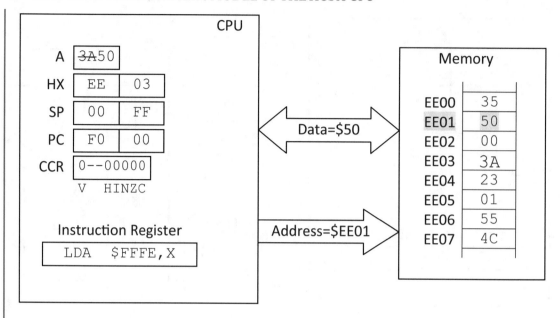

Figure 2.8: Illustration of 16 bit offset indexed addressing.

a) Hour into accumulator A
b) Minutes into accumulator A
c) Seconds into accumulator A

Solution:

a. The byte pointed to by HX is the first byte of the data structure, which is the hour byte. Thus, indexed addressing can be used

        LDA ,X

b. The next consecutive byte is the minutes. HX needs to be offset by 1 to generate the correct effective address. Thus, offset indexed addressing with offset 1 can be used

        LDA 1,X

c. The third byte is the seconds. Thus an offset of 2 is required

        LDA 2,X

Answer: a. LDA ,X   b. LDA 1,X   c. LDA 2,X

2.4.7   8B OFFSET INDEXED WITH POST-INCREMENT ADDRESSING

Offset indexed with post increment addressing has the same effective address as 8b offset indexed. After the effective address is computed, the CPU increments HX, just as in indexed post-increment addressing. This is illustrated in the Figure 2.9, which shows the CPU registers after the instruction executes. The HCS08 CPU has only one instruction that uses this mode: CBEQ. The mode is shown with the LDA Instruction in Figure 2.9 for clarity; the HCS08 CPU does not actually allow this opcode and addressing mode combination.

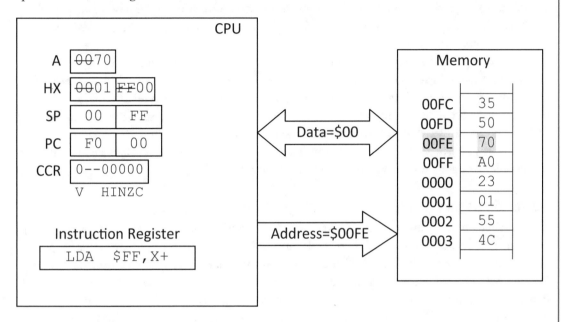

Figure 2.9: Operation of 8b Offset indexed with Post-Increment Addressing.

2.4.8   STACK POINTER OFFSET

Stack pointer offset mode is similar to indexed offset except the stack pointer SP is used as the base register. The effective address is computed as the content of SP plus an 8b or 16b offset. This mode is used to access items that have been placed on the stack; thus, discussion is deferred to the stack section later in this chapter.

2.4.9   RELATIVE ADDRESSING

Relative addressing is used only by branch instructions to compute the *target* of the branch. In this mode, a *signed* 8 bit offset is added to the program counter to compute the effective address. By the time the CPU computes the effective address, the PC has already been incremented to point to the instruction after the branch; thus, the effective address is computed as

$$\text{Effective Address} = \text{PC}_{\text{next}} + \text{sign\_extend(offset)}$$

where $\text{PC}_{\text{next}}=\text{PC}_{\text{branch}}+2$ and $\text{PC}_{\text{branch}}$ refers to the location of the branch in memory. Because the 8b offset is signed, it must be sign-extended by the CPU before computing the effective address.

Since the range of an 8b signed integer is -128 to +127, the effective address is limited to the range PC_{next}-128 to PC_{next}+127. When using an assembler, a programmer does not need to compute the offset; the programmer specifies the address of the instruction that the branch is to go to; the assembler performs the necessary computations to determine the offset. The programmer needs to remember, however, that the range of a branch target is limited to a small range centered about the branch instruction itself.

Encoding branch targets relative to the PC has advantages over using full 16b addresses to specify the targets. First, the machine code is smaller since an 8 bit offset is all that is needed to specify the operand (branch target). Second, because the target is specified relative to the PC of the branch instruction, it does not depend on where the program is located in memory. This means that a program can be placed anywhere in memory and the branch instructions will operate correctly. This property, known a relocatable code, allows an operating system to load programs into memory in any order without considering limitations on where they can be placed.

2.4.10   MEMORY-TO-MEMORY MOV ADDRESSING MODES

The remaining four addressing modes are compound addressing modes consisting of pairs of simple addressing modes discussed above. These modes are used exclusively in the only two memory operand instruction contained in the HCS08 CPU, the MOV instruction, which copies data from one memory location to another. Unlike the accumulating arithmetic and logical instructions, MOV must be a two-operand format instruction because it is necessary to specify both the source and destination locations of the copy. The HCS08 CPU limits the MOV instruction to four pairs of addressing modes, shown in the Table 2.1. Each of the modes is named using the format: "*Source Addressing Mode* to *Destination Addressing Mode*." For each of these modes, there are up to two effective addresses calculated by the CPU: a source effective address and a destination effective address. At least one of the source or destination always uses direct addressing, meaning that one of the operands must always reside within the direct page.

Table 2.1: Memory-to-Memory Move Addressing Modes.

Mode Name	Source Mode	Destination Mode	Syntax Example
Immediate to Direct	Immediate	Direct	MOV #$02,$F0
Direct to Direct	Direct	Direct	MOV $F0,$81
Indexed with Post-Increment to Direct	Indexed w/ Post increment	Direct	MOV X+,$FF
Direct to Indexed with Post-Increment	Direct	Indexed w/ Post-Increment	MOV $80,X+

Example 2.6. Provide an instruction (or instructions) to
   a) Copy a byte from address $0080 to address $00F4
   b) Initialize the byte at address $00E2 to AF_{16}
   c) Copy the byte from $FF34 to $0100

Solution:

a) The first case is a copy between two locations that can be reached by direct addressing. Thus a direct-direct move is appropriate. `MOV $80,$F4`

b) The source is a constant and the destination can be reached by a direct address. Thus an immediate-to-direct move can be used: `MOV #$AF,$E2`

c) In this case, neither address can be specified with direct addressing, so a `MOV` cannot be used. One way to accomplish this operation is:

```
LDA $FF34
STA $0100
```

Example 2.7. Provide an instruction (or instructions) to copy the byte at $00 four times, to locations $0034, $0035, $0036, and $0037

Solution: This operation is a copy from a direct address to 4 consecutive direct addresses. Two methods are possible using the `MOV` instruction: 4 direct-to-direct moves or 4 direct-to-index with post increment moves can be used.

```
MOV $00,$34        LDHX #$0034
MOV $00,$35        MOV $00,X+
MOV $00,$36        MOV $00,X+
MOV $00,$37        MOV $00,X+
                   MOV $00,X+
```

Although the sequence on the right has one extra assembly language instruction, it requires 1 byte less of machine code than the sequence on the left. This is because each `MOV` is smaller by 1 byte. On the other hand, a disadvantage is that it uses register `HX` while the left sequence does not.

2.5 DATA TRANSFER OPERATIONS

Data transfer instructions transfer data within the microcomputer. Most data transfer operations are copy operations, in which the source operand is copied to the destination. Some CPUs contain other forms of data transfer operations, such as exchange or swap operations.

Data transfers of the "copy" type generally can have one of 4 forms in a computer: loads, stores, register transfers and memory-to-memory moves. These are listed in Table 2.2. A load copies an operand from memory to a register. A store is the opposite of a load, with data copied from a register to a memory location. The terms load and store are specified relative to the CPU; thus, a CPU loads data from memory and stores data to memory. Register transfers copy the contents of one register to another. Memory moves copy the contents of one memory location to another.

Table 2.2: HCS08 Data Transfer Instruction Types and Direction of Data Flow.		
Data Transfer Operation Type	*Source Operand Location (from)*	*Destination Operand Location (to)*
Load	from Memory	to Register
Store	from Register	to Memory
Move	from Memory	to Memory
Register Transfer	from Register	to Register

Table 2.3 summarizes the data transfer instructions of the HCS08 CPU. In the table, *src* indicates a single source operand specified using one of the HCS08 CPU addressing modes; *dst* specifies a destination operand using one of the HCS08 CPU addressing modes. Not all addressing modes may be used with each instruction. Refer to the *HCS08 Family Reference Manual* for allowable addressing modes.

2.6 BIT MANIPULATION OPERATIONS

Bit manipulation instructions clear (change to (0) and set (change to (1) individual bits in memory or registers. These instructions are used frequently in manipulating the bits of I/O configuration registers as well as manipulating flags in the CCR and Boolean variables in memory. CCR flags can be used to some extent as Boolean variables; the difficulty is that other instructions, such as arithmetic instructions, may alter the flags as a side effect making it difficult to keep track of when the Boolean value stored in one of the flags has been altered. Despite this, however, flags find use as temporary Boolean variables as well as for passing Boolean data to/from subroutines. Logical instructions can also be used to set, clear or toggle multiple bits in memory or in the accumulator. These operations will be discussed in Section 2.6.

The bit manipulation instructions are summarized in Table 2.4. Instructions can clear or set bits in memory, registers or flags. For bit manipulation instructions that operate on bits in memory, only direct addressing is allowed as the addressing mode to identify the location in memory of the value containing the bit to be modified; the individual bit within the byte is explicitly specified as an operand in the instruction, with bit 7 indicating the msb and bit 0 the lsb.

Table 2.3: Data transfer instructions of the HCS08 CPU.

Instruction	Operation Type	Description
LDA src	Load	Loads an 8b operand from memory into accumulator A.
LDHX src	Load	Loads a 16b operand from memory into HX. The operand is big-endian: the MSB at the effective address is copied into H; the LSB at the effective address +1 is copied into X.
LDX src	Load	Loads an 8b operand from memory into X (H not affected).
MOV src,dst	Move	Copies the source (first) operand from memory to the destination location in memory.
NSA	Swap	Nibble swap accumulator: swaps the lower nibble and upper nibble of accumulator A.
STA dst	Store	Copies the byte in accumulator A to the memory location specified by the (destination) operand.
STHX dst	Store	Copies two bytes from HX to memory. The operand is stored in big endian byte order.
STX dst	Store	Copies the byte from X to memory location specified by the effective address of the destination operand.
TAP	Transfer	The contents of accumulator A are copied into the CCR.
TAX	Transfer	The contents of accumulator A are copied into X.
TPA	Transfer	The contents of the CCR are copied into accumulator A.
TXA	Transfer	The contents of X are copied into accumulator A.

Example 2.8.   Provide an instruction (or instructions) to
  a) Set the msb at memory location $00E6
  b) Clear bits 3 and 5 of location $0000

Solution:

  a) The memory address can be specified with direct addressing; thus a BSET instruction can accomplish this. The msb is bit 7, thus BSET 7,$E6 will set the msb of the value at effective address $00E6.

  b) Again, the address is direct. However, two bits need to be cleared, requiring 2 BCLR instructions

$$\text{BCLR 3,\$00}$$
$$\text{BCLR 5,\$00}$$

Table 2.4: Bit Manipulation instructions of the HCS08 CPU.

Instruction	Operation Type	Description
BCLR n,dir	Clear Bit in Memory	Bit number n at the memory location specified by the direct address is cleared.
BSET n,dir	Set Bit in Memory	Bit number n at the memory location specified by the direct address is set.
CLC	Clear Flag	The carry flag is cleared.
SEC	Set Flag	The carry flag is set.
CLI	Clear Flag	Clears the interrupt mask bit.
SEI	Set Flag	Sets the interrupt mask bit.

2.7   ARITHMETIC AND LOGIC OPERATIONS

Arithmetic and logical operations are the "bread-and-butter" of the CPU. They perform many of the operations necessary to achieve the data manipulation required by programs. As discussed above, arithmetic and logic operations in the HCS08 CPU are accumulative: the destination of the operation is also the left source operation. Results are accumulated in accumulator A or in memory. In this instruction group, simple operations such as clearing or negating a byte in memory have also been included; these simple operations are unary operations (single operand instructions) that are not accumulative.

The arithmetic and logic operations are summarized in Table 2.5. In the table, *opr* specifies the right source operand of the instruction; the left source operand is the same as the result, which is accumulated in a CPU register. The notation *dst* specifies the destination of a simple operation, such as clear. The operand *opr* or *dst* is specified using one of the HCS08 CPU addressing modes. The left arrow indicates the register on the left is replaced by the value of the expression on the right.

Example 2.9.   Provide a sequence of instructions to add +23 to index register HX.
Solution: AIX adds a signed 8b value to HX. Since +23 fits within an 8b signed code, this instruction can be used.
Answer: AIX #23.

Example 2.10.   Provide two methods for toggling all the bits in accumulator A.
Solution: COMA complements the bits of A and is the obvious and most efficient solution since it uses one byte of machine code. Another method is to exclusive-or A with #$FF, which requires 2 bytes of machine code.

Instruction	Operation Type	Operation	Description
ADC opr	Arithmetic	$A \leftarrow A + opr + C$	Operand+C flag is added to A. C-flag is zero-extended.
ADD opr	Arithmetic	$A \leftarrow A + opr$	Operand is added to accumulator A
AIX opr	Arithmetic	$HX \leftarrow HX + opr$	Signed, 8b immediate operand is sign-extended and added to HX
AND opr	Logical	$A \leftarrow A \wedge opr$	Operand is logical ANDed to A
CLR dst	Simple	$dst \leftarrow 0$	Clears the memory operand
CLRA	Simple	$A \leftarrow 0$	Clear A
CLRH	Simple	$H \leftarrow 0$	Clears H (upper byte of HX)
CLRX	Simple	$X \leftarrow 0$	Clears X (lower byte of HX)
COM dst	Simple	$dst \leftarrow \overline{dst}$	Complements the memory operand
COMA	Simple	$A \leftarrow \overline{A}$	Complements A
COMX	Simple	$X \leftarrow \overline{X}$	Complements X
DAA	Simple		Corrects the result in accumulator A after BCD addition
DEC dst	Arithmetic	$dst \leftarrow dst - 1$	Decrements the memory operand
DECA	Arithmetic	$A \leftarrow A - 1$	Decrements A
DECX	Arithmetic	$X \leftarrow X - 1$	Decrement X
DIV	Arithmetic	$A \leftarrow \frac{HA}{X}$	Unsigned division (H:A) $\div$X; quotient in A, remainder in H
EOR opr	Logical	$A \leftarrow A \oplus opr$	Operand is exclusive-ORed to A
INC dst	Arithmetic	$dst \leftarrow dst + 1$	Increment memory operand
INCA	Arithmetic	$A \leftarrow A + 1$	Increment A
INCX	Arithmetic	$X \leftarrow X + 1$	Increment X
MUL	Arithmetic	$XA \leftarrow X \times A$	Unsigned multiplication X $\times$ A; product placed in X:A
NEG dst	Simple	$dst \leftarrow -(dst)$	Twos complement the memory operand
NEGA	Simple	$A \leftarrow -(A)$	Twos complement A
NEGX	Simple	$X \leftarrow -(X)$	Twos complement X
ORA opr	Logical	$A \leftarrow A \vee opr$	Operand is logical ORed to A
SBC opr	Arithmetic	$A \leftarrow A - (opr + C)$	(Operand+C) subtracted from A; C is zero-extended
SUB opr	Arithmetic	$A \leftarrow A - opr$	Operand is subtracted from A

Table 2.5: Arithmetic and Logical instructions of the HCS08 CPU.

Answer: EOR #$FF or COMA.

Example 2.11.   Provide a sequence of instructions to compute X/17, where X is an 8 bit unsigned value in memory at $E19F. The quotient is to be stored at $0043 and the remainder at $0078.

Solution: The HCS08 CPU division instruction divides the unsigned 16b dividend in H:A by the divisor in X. Because the problem specifies an unsigned 8b dividend, it must be zero extended. The division can be set up with

```
        LDHX #0       ; clear H, for zero extension
        LDA $E19F     ; load dividend into A
        LDX #17       ; load divisor into X
        DIV           ; divide H:A / X
```

The quotient after division is in A and the remainder in X. Thus, to store the result we need

```
        STA     $43 ; store quotient
        STX     $78 ; store remainder
```

Answer:

```
    LDHX #0         ; set H to 0 for zero extension
    LDA $E19F       ; load dividend into A
    LDX #17         ; load divisor into X
    DIV             ; divide H:A / X
    STA     $43     ; store quotient
    STX     $78     ; store remainder
```

Example 2.12.   Provide a sequence of instructions to add +23 to index register X without altering H.

Solution: AIX adds a signed 8b value to HX; thus, it cannot be used here. There is no instruction to add to X. By transferring the operand temporarily to A, the addition can be performed and the result moved back into X. Note that the original value in A is lost.

Answer:

```
        TXA         ; copy value from X to A
        ADD #23     ; add 23 to it
        TAX         ; move it back into X
```

2.7.1   MASKING

It is often necessary to set, clear or toggle multiple bits in an operand. For these operations, bit-manipulation instructions are not well-suited because they affect only single bits; furthermore, they cannot perform a toggle operation. Logical AND, OR and EOR operations with an immediate can be used, however, to achieve the desired result. For example, to set bits 3, 5, and 7 in accumulator A and not change the other bits, one can use the instruction

```
ORA #%10101000; set only bits 3, 5 and 7 in A
```

The immediate operand in the OR instruction is referred to as a *mask*. The logical OR is performed bitwise; that is, each bit in A is ORed with its corresponding bit from the mask. Because 0 is the identity element with respect to logical OR, each bit of accumulator A that correspond to a '0' in the mask is left unaltered; each bit in accumulator A that corresponds to a '1' in the mask is set. Because the mask in this case is logical ORed with the operand, it is referred to as an OR-mask. An OR mask can be used to set individual bits in an operand.

There are three types of masking operations. An OR-mask is used to set bits in the operand; an AND-mask is used to clear bits in the operand; and an EOR-mask is used to toggle bits in the operand. Because the identity element for OR and EOR is 0, OR-masks and EOR-masks operate on the bits of the operand for which the mask bit are 1. AND masks operate on bits of the operand for which the corresponding mask bits are 0 (the identity element is 1). This is summarized in Table 2.6.

Table 2.6: Masking Operations.		
Mask Type	*Function*	*Description of Masking Operation*
AND-mask	Clear Bits	For each '0' in the mask, the corresponding bit of the operand is cleared; all other bits of the operand remain unchanged.
OR-mask	Set Bits	For each '1' in the mask, the corresponding bit of the operand is set; all other operand bits remain unchanged.
EOR-mask	Toggle Bits	For each '1' in the mask, the corresponding bit of the operand is toggled; all other bits remain unchanged.

Example 2.13.   Provide instruction(s) to toggle bits 5, 6, and 7 in memory location $44.
Solution: A toggle operation on multiple bits uses an EOR-mask. The HCS08 EOR instruction exclusive-ORs it operand with the contents of A. Thus, we can load the value from location $44, mask it, and store it back:

```
LDA    $44          ; get operand into A
EOR    #%11100000   ; toggle bits 7,6,and 5
STA    $44          ; write back result
```

An alternative is to put the mask into A, EOR the contents of $44 with it, and write back the result, as in

```
LDA    #%11100000  ;load mask into A
EOR    $44          ;exclusive-or operand with mask
STA    $44          ;write back result
```

The two sequences leave all registers and memory equally affected.

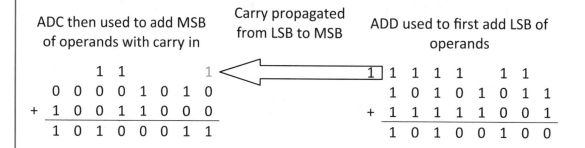

Figure 2.10: Using ADC to achieve extended addition on the HCS08 CPU.

2.7.2   EXTENDED ARITHMETIC

The ADC (add with carry) and SBC (subtract with borrow) instructions can be used to implement instruction sequences that accomplish addition and subtraction on operands that are larger than 8 bits. Because the HCS08 CPU only supports 8b addition and subtraction operations, operations on these wider operands are called extended arithmetic operations.

When adding 8b operands, the carry bits percolate through the operation from least-to-most significant bit; after the addition the carry flag holds the carry out of the msb. This carry out can be added to the lsb of the addition of the next byte, as shown in Figure 2.10, to extend the addition. This extension is accomplished with the ADC instruction, which performs an addition of the value in accumulator A, the operand and the (zero-extended) carry flag. In this way, the carry can be percolated through the bytes of the addition to create addition operations of any width. An extended subtraction can also be achieved in this way using the SBC instruction.

Example 2.14.   An instruction sequence to increment the 16b unsigned value at address $0080 would first increment the LSB of the 16b operand (at address $0081), then add the carry to the MSB of the operand (at address $0080), as shown below.

```
INC  $0081      ;Add 1 to the LSB
ADC  #0         ;Add carry to MSB
```

Example 2.15. Provide instruction(s) to add the two 16 bit signed operands at $FFC0 and $EE00 and store the 16b result at $0080.

Solution: Extended addition is required. The LSB of the source operands are at $FFC1 and $EE01 – these need to be added first.

```
LDA  $FFC1      ; load LSB of one operand into A
ADD  $EE01      ; add LSB of second operand
STA  $81        ; store to LSB of result
```

Note that the store does not alter the carry flag; this is important to preserve the carry out of the LSB. Then, the MSB of the operands at $FFC0 and $EE00 must be added, taking into account the carry from the addition of the LSBs.

```
LDA  $FFC0      ; load MSB of first operand
ADC  $EE00      ; add MSB of second operand, with carry
STA  $80        ; store result
```

Answer:

```
LDA  $FFC1 ; load LSB of one operand into A
ADD  $EE01 ; add LSB of second operand
STA  $81        ; store to LSB of result
LDA  $FFC0      ; load MSB of first operand
ADC  $EE00      ; add MSB of second operand +carry
STA  $80        ; store result
```

2.8 SHIFT AND ROTATE

Shift and rotate instructions shift the bits of the destination operand left or right by one bit position each time they are executed. There are no instructions in the HCS08 CPU instruction set to shift or rotate multiple times.

The right and left shift operations are illustrated in Figure 2.11. Shift instructions move all of the bits in the operand to the right or to the left by one position. Each time a shift occurs, a bit is shifted out of the msb or lsb (depending on the direction of the shift) and that bit is placed in the carry flag. Each time a shift occurs, one bit position becomes vacated: a shift right leaves the msb

Logical Shift Right Operation

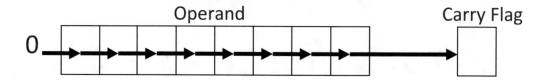

Arithmetic Shift Right Operation

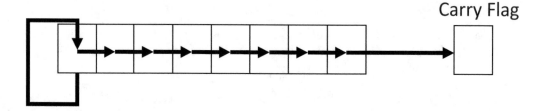

Shift Left Operation (Logical or Arithmetic)

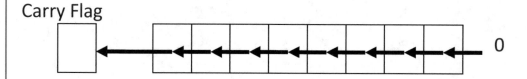

Figure 2.11: Operation of the HCS08 CPU shift instructions.

"vacant" and a shift left leaves the lsb "vacant." The value of the vacant bit defines the type of shift. Left shift operations always clear the vacant lsb (a 0 is shifted in).

Right shift operations come in two forms: logical and arithmetic. A logical right shift operation shifts a zero into the msb. A logical right shift is equivalent to unsigned division by 2. Arithmetic right shift operations do not change the value of the msb – this has the effect of sign-extending the operand in the register. An arithmetic right shift operation is equivalent to signed division by 2. In both cases, the remainder after the division is in the carry flag. Some CPUs, like the HCS08 CPU, define both arithmetic and logical left shift mnemonics; however, these mnemonics map to the same machine code instruction and are identical; the two names are provided for programmer convenience.

Rotate operations are similar to shifts, but the bit shifted into the msb or lsb comes from the carry flag. This is illustrated in Figure 2.12. Thus, the rotate operations "rotate through" the carry

Rotate Right Operation

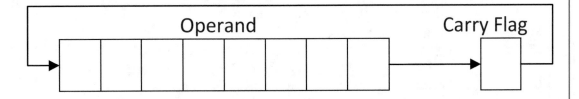

Rotate Left Operation

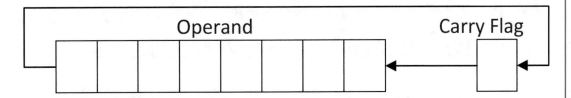

Figure 2.12: Operation of the HCS08 CPU rotate instructions.

flag to allow testing of each bit as it is moved out. Thus, a total of 9 repeated rotate operations restore the operand back to its original value. Rotate operations are logical – they have no inherent arithmetic meaning.

The shift and rotate instructions of the HCS08 CPU are summarized in Table 2.7. As before, *dst* specifies a destination operand; a memory location to be shifted or rotated.

Example 2.16.   Provide instructions to divide the 8 bit unsigned value in memory at $00E1 by 4.
Solution: As seen in a previous example, division requires several instructions to transfer the operands to the correct registers. Division by a power of 2, however, is equivalent to a shift right by 2. Because the division is unsigned, a logical shift right is needed to zero extend the quotient.

Table 2.7: Rotate and shift instructions of the HCS08 CPU.

Instruction	Operation Type	Description
ASL dst	Arithmetic	Left shift of destination operand. Identical to LSL.
ASLA	Arithmetic	Left shift of accumulator A. Identical to LSLA.
ASLX	Arithmetic	Left shift of register X Identical to LSLX.
ASR dst	Arithmetic	Right shift of destination operand.
ASRA	Arithmetic	Right shift of accumulator A.
ASRX	Arithmetic	Right shift of register X
LSL dst	Logical	Left shift of destination operand. Identical to ASL.
LSLA	Logical	Left shift of accumulator A. Identical to ASLA.
LSLX	Logical	Left shift of register X Identical to ASLX.
LSR dst	Logical	Right shift of destination operand.
LSRA	Logical	Right shift of accumulator A.
LSRX	Logical	Right shift of register X
ROL dst	Logical	Left rotate of destination operand.
ROLA	Logical	Left rotate of accumulator A.
ROLX	Logical	Left rotate of register X
ROR dst	Logical	Right rotate of destination operand.
RORA	Logical	Right rotate of accumulator A.
RORX	Logical	Right rotate of register X

```
LSR $E1    ;divide operand at $00E1 by 2
LSR $E1    ;divide operand at $00E1 by 2 (again)
```

Example 2.17.   Repeat the previous example, assuming the operand is signed.
Solution: Because the division is signed, an arithmetic shift right is needed to ensure proper sign extension.

```
ASR $E1    ;divide operand at $00E1 by 2
ASR $E1    ;divide operand at $00E1 by 2
```

Example 2.18. Provide a sequence of instructions to Z=3V, where Z the 8 bit unsigned value in memory at $00E1 and V is an 8 bit variable at $0034.

Solution: MUL takes several instructions to set up the data and requires 5 clock cycles to execute. A slightly more effective solution is to compute 3V as 2V+V, which requires a shift and an add.

```
LDA    $34 ;load V into A
LSLA       ;multiply by 2
ADD    $34 ; add V
STA    $E1 ; store result to Z
```

Example 2.19. Provide a sequence of instructions to multiply the 16 bit unsigned value in memory at $00E1 by 4.

Solution: Again, the multiply is by a small power of 2, so shifts are appropriate. Since the operand is 16b and only 8bit shifts are possible, 2 shifts are necessary. A shift left of a 16b operand is equivalent to a shift left of the least significant byte followed by a rotate left of the most significant byte. Why a rotate left? The bit shifted out of the msb of the lower byte will be rotated into the lsb of the upper byte. This sequence is repeated twice to get a multiply by 4. Also, recall that with big-endian byte ordering, the lower byte is at $00E2.

```
LSL $E2    ; shift left low byte; save msb in carry
ROL $E1    ; shift left high byte, carry into lsb
LSL $E2    ; repeat
ROL $E1
```

Example 2.20. Provide a sequence of instructions to compute V=3V+2U, where V is a signed 8b value in memory whose address is in HX (i.e., HX points to V), and U is another 8b signed value at $88. Ignore overflow.

Solution: Shifts and adds can be used to perform the two multiplies. By rearranging the expression as V+2(V+U), the sequence can be accomplished with an add (V+U), shift, and an ADD.

```
LDA    ,X     ; load V into A
ADD    $88    ; add U
LSLA          ; multiply by 2 (ignore overflow)
```

```
ADD    ,X   ; add V
STA    ,X   ; store result back to V
```

2.9   STACK OPERATIONS

The stack is a region of memory used as a Last-In-First-Out queue. A stack can be thought of as a data structure that can grow to arbitrary size (within the limits of free memory available in the microcomputer for the stack). There are many instances in computing where it cannot be predicted before a program runs exactly how much memory will be required to implement an algorithm. Because stacks can grow "arbitrarily," they are effective data structures to use in these situations; their elasticity allows them to grow to meet program demand. Of course, practical limitations place restrictions on the growth of the stack; programs that exceed these limits typically crash. CPUs generally maintain a CPU stack that serves the memory needs of subroutines and interrupts. The CPU stack can also be used for temporary program data storage. The CPU stack is supported by the existence of the stack pointer register SP and stack manipulation instructions.

A stack structure in memory can be thought of like a stack of books or stack of coins; items (data) can be placed on top of the stack and removed from the top of the stack. In a computer stack, items cannot be removed from the middle or bottom of the stack. Placing data on the stack is called a *Push* and removing data is called a *Pull*. Over the lifetime of a program, the stack grows and shrinks, meeting the demands of the program. Figure 2.13 illustrates the elastic "push-pull" nature of the stack. Data values pushed onto the stack cover existing stacked data. The value at the top of the stack is the only one available to be removed; a value that is removed can be used by an instruction as an operand. Although data cannot be removed from the middle of the stack, if its position is known a program can index into the stack to read or alter its value; this is the reason for having the stack pointer indexed addressing mode. Because it is possible to access data on the stack in this way, the stack can be used for program data storage.

Because the stack is an elastic structure, programs require a mechanism to track the location of the top of the stack. The stack pointer is a special register whose exclusive purpose is to keep track of the top of the stack. The stack pointer automatically follows the top of the stack as items are pushed and pulled. The bottom of the stack is anchored in memory – it starts at an initial location and does not change over the lifetime of a program.

Although the stack is visualized as growing upwards, generally the stack grows towards lower memory addresses. This is illustrated in Figure 2.14, which shows a RAM memory map of a computer's address space. A typical program organization has the machine code placed at the bottom of memory and uses the area above the program for its data. The end of the data area, however, provides no fixed reference point to which the bottom of the stack can be anchored; however, the top of RAM is a fixed, known starting point that can be used to anchor the bottom of the stack.

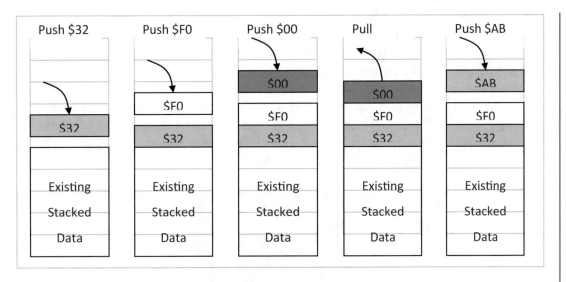

Figure 2.13: Illustration of stack push and pull operations.

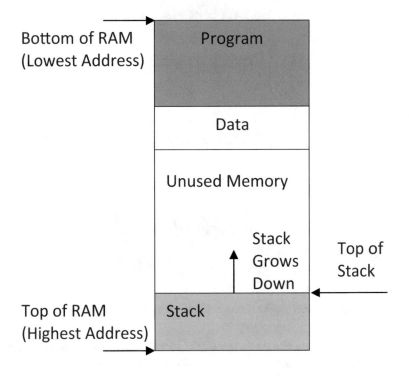

Figure 2.14: CPU memory map showing stack area.

The HCS08 CPU stack grows downward in memory; because of this, the stack pointer is typically initialized to point to the top of RAM. A push operation places the data on the top of the stack and decrements the stack pointer by the size of the data pushed, effectively keeping SP pointing to the (new) top of the stack. As a result of decrementing SP after each push, SP always points to the next empty location at the top of the stack. A pull operation decrements the stack pointer to point to the last item on the stack, then copies the data off the top of the stack. Figure 2.15 illustrates the relationship between the stack pointer and the top of the stack. The "Empty Slot" pointed to by SP is simply an uninitialized memory location. The existing stacked data starts from address SP+1; thus, the last item stacked is always at the lowest address.

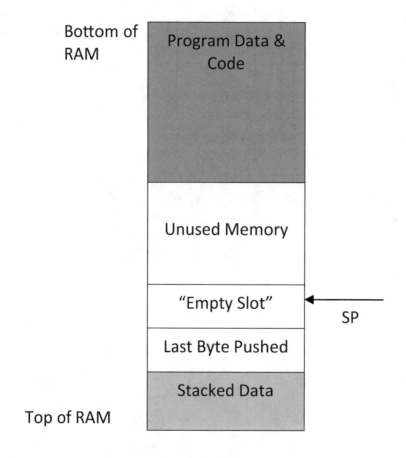

Figure 2.15: Relationship between stack pointer and top of stack.

Once a value is pulled from the stack, the RAM location that contained that value cannot be expected to retain a copy of the value; even immediately after it is pulled. This is because interrupts, which can occur at any time (even between two instructions), use the stack for data storage. Thus, the RAM in the region below the stack pointer in memory should always be considered volatile: its contents should never be expected to remain constant even between two instructions. Assuming the value pulled off the stack still exists in RAM after the pull is a common mistake made by programmers.

The stack has no identifiable "landmarks" other than SP. Once the stack is in use, it is no longer possible to identify the bottom of the stack. Yet, this structure can be shared by several uncoordinated software entities (subroutines, interrupts, etc.) at any given time. Despite the lack of coordination among the entities sharing the stack, no information other than SP is required to maintain the stack. The stack works because a balanced stack discipline is followed; upon exiting, each software entity that uses the stack must leave the stack structure exactly as it was found upon entry; though values can be modified, no additional items should be left on top of the stack and nothing extra should be removed. For this discipline to work, every software section should balance stack operations: for each push operation, there should be a corresponding pull operation to clean the stack. Unbalanced stack operations will lead to erroneous program behavior.

The HCS08 CPU contains instructions to push and pull data as well as to manipulate the stack pointer. These instructions are summarized in Table 2.8. In addition, SP indexed addressing can be used with other instructions to access data on the stack.

When the HCS08 CPU is first powered up, the stack pointer is initialized to $00FF. This is to maintain compatibility with older HC05 processors. If $00FF is not the top of RAM on a specific HCS08 family member, one of the first instructions executed by the system after power-up should be to initialize the stack pointer to point to the correct location at the top of RAM. This is necessary because the stack is also used for interrupt processing.

Example 2.21.   Provide a sequence of instructions to swap the byte in X with the byte in A.
Solution: HCS08 CPU has instructions to transfer bytes, but not swap them. A general swap operation on values A and X is done by copying A to a temporary location, Copying X to A, and finally copying from the temporary location to X. The temporary location is required to hold one of the operands since the first copy operation will, by necessity, make both locations A and X the same. First, the value in A is pushed to the stack

> PSHA

Then, the value in X is copied to A

> TXA

Finally, the temporary copy of the original A value on the stack can be pulled into X

> PULX

Answer:

```
PSHA   ; save A on stack (temp=A)
TXA    ; copy X to A (A=X)
PULX   ; pull original A into X (X=temp)
```

Table 2.8: Stack manipulation instructions of the HCS08 CPU.

Instruction	Operation Type	Description
AIS opr	General Stack	Sign-extend and add the 8b signed immediate opr to SP. Used to create uninitialized space on top of stack or to delete multiple items from the stack top.
PSHA	Push	The value in accumulator A is copied to the top of the stack. The stack pointer is then adjusted by -1.
PSHH	Push	The value in register H is copied to the top of the stack. The stack pointer is then adjusted by -1.
PSHX	Push	The value in register X is copied to the top of the stack. The stack pointer is then adjusted by -1.
PULA	Pull	The stack pointer is adjusted by +1 to point to the top data byte on the stack. The byte is then copied into Accumulator A. Immediately after the instruction, the data byte pointed to by SP is considered undefined.
PULH	Pull	The stack pointer is adjusted by +1 to point to the top data byte on the stack. The byte is then copied into register H. Immediately after the instruction, the data byte pointed to by SP is considered undefined.
PULX	Pull	The stack pointer is adjusted by +1 to point to the top data byte on the stack. The byte is then copied into register X. Immediately after the instruction, the data byte pointed to by SP is considered undefined.
RSP	General Stack	Resets the LSB of SP to FF, pointing it again to the top of RAM. This instruction does not alter MSB of SP (i.e., top of RAM assumed to be $00FF). Effectively removes everything from stack and should be used with care.
TSX	General Stack	Copies SP+1 to index register HX. This makes HX point to top valid data byte on Stack.
TXS	General Stack	Copies HX-1 to SP. Opposite of TSX. Assuming HX points to the element that is to be the top item on the stack, SP is updated.

Example 2.22. HX points to an 8b operand (P) in memory. Compute the logic operation (AA_{16} AND P) OR (55_{16} AND NOT P). The result should be kept in accumulator A and HX should continue to point to P after the sequence executes.

Solution: The two AND expressions can be computed separately and then logical-ORed. The first expression can be accomplished with

```
LDA    ,X  ; loads operand P
AND    #$AA; AND it with $AA
```

The result of the first expression needs to be saved somewhere because A is needed to compute the second expression; register HX is in use, so neither H nor X can be used to temporarily hold the contents. It can, however, be pushed onto the stack:

```
PSHA     ; temporarily store result on stack
```

Then, the second expression can be computed

```
LDA    ,X     ; load operand P
COMA          ; complement it
AND    #$55   ; AND it with 55h
```

Now, the expression in accumulator A needs to be ORed with the expression on the stack. SP is indexed with -1 to get the value of the byte at the top of the stack.

```
ORA    1,SP ;OR with saved expression on top of stack
```

Now, the result of the first expression must be cleaned from the stack. Because all registers are in use, instead of pulling the value SP can be incremented by 1 to accomplish this.

```
AIS    #01    ;remove data from stack
```

Answer:

```
LDA    ,X     ; loads operand P
AND    #$AA   ; AND it with $AA
PSHA          ; temporarily store result on stack
LDA    ,X     ; load operand P
COMA          ; complement it
AND    #$55   ; AND it with 55h
ORA    1,SP   ; OR with saved expression on top of stack
AIS    #01    ; remove data from stack
```

2.10   COMPARE AND BRANCH OPERATIONS

Recall that the CPU executes instructions one after another in sequential fashion – the next instruction is always found in memory immediately after the current instruction. Restricting the CPU to follow such a straight-line path through memory means that the CPU would always perform exactly the same set of computations. Most practical algorithms, however, require decision making including iteration and other forms of program flow control changes. Implementing these flow changes requires the CPU to leave the current linear instruction sequence and begin following another. At a minimum, supporting control flow structures requires the CPU to have the ability to conditionally execute an instruction and the ability to alter instruction sequencing to allow execution of another part of a program. Most CPUs contain branch and jump operations for this purpose. Figure 2.16 illustrates one need for flow control changes. The flowchart of a simple decision type algorithm is shown; the blue and red lines are the two possible execution sequences that can result from the algorithm, depending on the value of Y during execution. Because of the linear nature of the computer memory, however, the instructions in memory must be strictly ordered. Thus, the structure of the flowchart must be "flattened" into a linear sequence of machine code instructions, as shown on the right side of the figure. The black lines in memory indicate normal CPU instruction flow. The red and blue lines indicate the deviations from this flow that are necessary to implement the two execution sequences shown in the flow chart.

Branch and Jump instructions are control flow altering instructions – they cause the CPU to change from executing one part of a program to another. Because the CPU uses the program counter (PC) to identify the next instruction to execute, branch and jump instructions operate by changing the value in the program counter. Encoded in the branch or jump instruction is the target – the address of the instruction that the CPU should execute next.

Jump instructions are unconditional – that is, the jump is always taken. The next instruction executed after a jump is always the instruction at the target. Branch instructions, on the other hand, have a condition associated with them. When the CPU executes the branch instruction, it evaluates the condition and branches to the target only if the condition is true; in this case, the branch is said to be taken. If the branch condition is false, the CPU continues execution from the instruction immediately following the branch in memory; the branch in this case is not taken. The not taken path of a branch is called the fall-through path and when a branch condition is false, the CPU is said to have fallen-through the branch.

To program a branch in assembly language, it is necessary to identify the instruction that is to be the target of the branch or jump. This is accomplished using labels at the start of each line of assembly code that is to be the target of a branch or jump. The colon ':' after the label separates it from the rest of the line and is not part of the label. As an example, the assembly language code below illustrates the use of labels. Because an assembler needs to differentiate between labels and opcodes, most assume that any identifier starting in the first column of a line is a label. Therefore, lines without labels must not begin in the first column.

```
LOOP:   ADD    #3      ;add 3 to A
        BLE    LOOP    ;repeat if A<=0
```

Like many CPUs with condition codes (flags), the HCS08 CPU branch conditions are based on the flags, which are set after most arithmetic and logic instructions. In addition, the CPU has some compare and test instructions whose main purpose is to set the flags; these instructions perform an arithmetic or logic operation without saving the result. The flags are set after the operation, however, to be used by branch conditions. For example, a compare (CMP) instruction subtracts its operands (without saving the results). If the zero flag is set after the compare, it indicates that the operands were equal (the difference is zero). Other flag conditions can be used to test a variety of operand relationships, such as greater-than-or-equal. Each flag condition has a complement; for example BMI branches if N=1; the complement branch, BPL, branches if N=0. Note that this is the largest

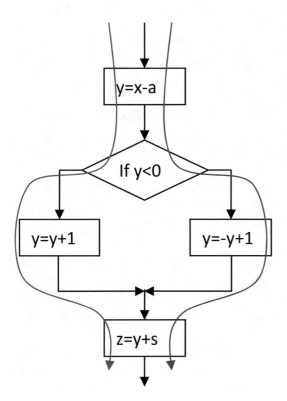
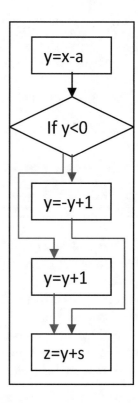

Figure 2.16: Using flow control for conditional execution. Flowchart on the left is restructured as an ordered sequence of operations on the right.

group of instructions in the instruction set, indicating the importance of control instructions in programming.

Table 2.9 lists the compare and test instructions of the HCS08 CPU. These instructions have no effect on processor registers, except for flags in the condition code register. Test type operations compare the operand with zero. Compare operations compare the value in the specified accumulator or index register with the operand. The BIT operation performs a logical AND of accumulator A and the operand; the result is not stored (accumulator A is unaffected) but the zero and negative flags are set based on the value of the result.

Table 2.9: Compare and test instructions of the HCS08 CPU.

Instruction	Operation Type	Description
BIT opr	Other	ANDs the operand with the value in accumulator A, discarding the result. Z and N are set based on the result of the AND operation.
CMP opr	Compare	Subtracts the operand from accumulator A, discarding the result. The V, N, Z, and C flags are affected.
CPHX opr	Compare	Compares (subtracts) the 16b operand from HX and discards the result. The V, N, Z, and C flags are affected.
CPX opr	Compare	Compares (subtracts) the 8b operand from X and discards the result. The V, N, Z, and C flags are affected.
TST opr	Test	The Z and N flags are set or cleared to reflect the value of the operand (zero or negative).
TSTA	Test	The Z and N flags are set or cleared to reflect the value in A (zero or negative).
TSTX	Test	The Z and N flags are set or cleared to reflect the value in X.

A summary of the HCS08 CPU branch and jump instructions are provided in Table 2.10. The operand rel refers to a branch target specified using Relative Addressing (Section 2.4.9). The conditions causing the CPU to branch are described, but the interpretation of these conditions is covered in the Section 2.10.1. The one jump instruction is unconditional and can use direct, extended or offset addressing.

Table 2.11 lists the HCS08 CPU instructions that perform two operations: a compare/test and a branch. These compound control transfer instructions usually save machine code and/or clock cycles compared to their equivalent two instruction sequences.

Example 2.23. Provide a sequence of instructions that will count the number of bits that are set in the byte at address $80. The count should be in accumulator A.

Table 2.10: Control transfer instructions of the HCS08 CPU.

Instruction	Operation Type	Description
BCS rel	Branch	Branch if carry set. Taken if (C=1)
BCC rel	Branch	Branch if carry clear. Taken if (C=0)
BEQ rel	Branch	Branch is equal. Taken if (Z=1).
BGE rel	Branch	Branch if greater than or equal (signed). Taken if N=V=0 or N=V=1 (i.e., N⊕V=0).
BGT rel	Branch	Branch if greater than (signed). Taken if Z=0 and N⊕V=0.
BHCC rel	Branch	Branch if half-carry flag clear. Taken if H=0.
BHCS rel	Branch	Branch if half-carry flag set. Taken if H=1.
BHI rel	Branch	Branch if higher (greater than, unsigned). Taken if C=Z=0.
BHS rel	Branch	Branch if higher or same (greater than or equal, unsigned). Taken if C= 0. Alternate mnemonic for BCC.
BIH rel	Branch	Branch if interrupt high. Taken if the interrupt pin (IRQ) has a high logic level applied.
BIL rel	Branch	Branch if interrupt low. Taken if the interrupt pin (IRQ) has a low-logic level applied.
BLE rel	Branch	Branch if less than or equal (signed). Taken if Z=1 or N⊕V=1.
BLO rel	Branch	Branch if lower than (unsigned). Taken if C=1. Alternate mnemonic for BCS.
BLS rel	Branch	Branch if low or same (unsigned). Taken if C=1 or Z=1.
BLT rel	Branch	Branch if less than (signed). Taken if N⊕V=1.
BMC rel	Branch	Branch if interrupt mask clear. Taken if I=0.
BMI rel	Branch	Branch if minus. Taken if N=1.
BMS rel	Branch	Branch if interrupt mask set. Taken if I=1.
BNE rel	Branch	Branch if not equal. Taken if Z=0.
BPL rel	Branch	Branch if plus. Taken if N=0.
BRA rel	Branch	Branch always. Unconditional.
BRN rel	Branch	Branch Never. Never branches (useful for debugging).
JMP opr	Jump	Jumps to the address specified. Direct, extended, and (offset) indexed are allowed modes.

Table 2.11: Compound-branch instructions of the HCS08 CPU.

Instruction	Operation Type	Description
BRCLR n,dir,rel	Test and Branch	Branch if bit clear. Branches if bit n at direct address dir is clear.
BRSET n,dir,rel	Test and Branch	Branch if bit set. Branches if bit n at direct address dir is set.
CBEQ opr,rel	Compare and Branch	Compare and branch if equal. If the operand is equal to the value in accumulator A, the branch is taken.

Solution: We can load the byte from $80 into index X. X can then be repeatedly shifted; the bit shifted out into the carry flag can be accumulated in A with ADC. This is repeated 8 times.

```
CLRA        ;initialize A to 0
LDX $80     ;load operand into X
LSLX        ;shift msb into C
ADC #$00    ;accumulate the bit in A
LSLX        ;repeat 7 more times
ADC #$00
LSLX
ADC #$00
LSLX
ADC #$00
LSLX
ADC #$00
LSLX
ADC #$00
LSLX
ADC #$00
LSLX
ADC #$00
```

However, repetition is often a sign that a loop can be used. In this example, at the end of the operation, 8 zeros have been shifted into X; therefore, repeating until X is equal to zero provides a simple method to loop. TSTX can be used to check if X is zero; BNE will branch if it is not zero, allowing to repeat the loop.

```
        CLRA        ;initialize A to 0
        LDX $80     ;load operand into X
LOOP:   LSLX        ; shift out a bit
        ADC #0      ; accumulate it
        TSTX        ; check if X is zero
        BNE LOOP    ; repeat loop if not
```

2.10.1 INTERPRETATION OF BRANCH CONDITIONS

The branch conditions are based on the CPU flags, which in turn are set or cleared by arithmetic, logic, test, compare, load and store operations. Depending on the type of operation that set or cleared the flags last as well as the type of operands (for example signed or unsigned), the branch conditions can have different meanings. Some branch conditions don't have meaning for all operation or operand types. This section is intended to provide guidelines for choosing appropriate branch types, not to be a comprehensive set of rules. It is up to the programmer to determine the appropriate branch to use in each situation.

Simple branches are based on the value of single flags. The meaning of the condition is simply the state of the flag being tested. Some might have other significance in certain situations; for example, C=1 indicates unsigned overflow after an unsigned arithmetic operation. The simple branch conditions are listed in the Table 2.12. Each branch is paired with its complementary branch.

Table 2.12: Simple branch conditions of the HCS08 CPU.			
Branch Mnemonic	*Condition*	*Complementary Mnemonic*	*Condition*
CS (carry set)	C=1	CC (carry clear)	C=0
EQ (equal)	Z=1	NE (not equal)	Z=0
HCC (half-carry clear)	H=0	HCS (half carry set)	H=1
IH (IRQ pin high)	IRQ=1	IL (IRQ pin low)	IRQ=0
MC (mask clear)	I=0	MS (mask set)	I=1
A (always)	1	N (never)	0

Signed arithmetic branches have meaning only after an instruction has performed an operation on signed operands, including load, store, compare, test, arithmetic and logic operations. The branch conditions that have significance for signed operands are listed in Table 2.13. Branch conditions not listed in this table generally are meaningless for signed operands and should not be used (for example, BCS would have no meaning after adding two signed operands).

Table 2.13: Branch conditions that can be tested after a signed arithmetic operations.			
Branch Mnemonic	*Condition*	*Complementary Mnemonic*	*Condition*
EQ (equal)	Z=1	NE (not equal)	Z=0
GE (greater than or equal)	$N \oplus V = 0$	LT (less than)	$N \oplus V = 1$
GT (greater than)	Z=0 and $N \oplus V = 0$	LE (less than or equal)	Z=1 or $N \oplus V = 1$
MI (minus)	N=1	PL (plus)	N=1

The interpretation of the branch mnemonics after each type of operation is summarized in Table 2.14. Where a condition does not include the overflow flag, the overflow is not accounted for in the test. Thus, the actual condition could be wrong (for example, EQ is true but there was overflow, meaning the actual result may not be zero). When a condition includes V, it accounts for overflow, meaning the branch outcome reflects the true value of the result as if the overflow had never occurred (even if the result in the accumulator is wrong). Load, store and test operations can't overflow and always clear the V flag; the condition for BGE in this case, for example, reduces from $N \oplus V=0$ to $N=0$, thus BGE, after a load or store, simply means the loaded operand is ≥ 0.

Table 2.14: Summary of branch conditions after signed data transfer, arithmetic, and compare instructions.

Branch Mnemonic	Meaning after Load/Store or Test of Operand	Meaning after Arithmetic operation (including ASL)	Meaning after a Compare of operand to A
EQ/	Operand=0/	Result=0/	A=Operand/
NE	Operand≠0	Result≠0	A≠Operand
LT/	Operand<0/	Result<0/	A<Operand/
GE	Operand≥0	Result≥0	A≥Operand
GT/	Operand>0/	Result>0/	A>Operand/
LE	Operand≤0	Result≤0	A≤Operand
MI/	Operand<0/	Result<0/	A<Operand/
PL	Operand>0	Result>0	A>Operand

Unsigned arithmetic branches have meaning only for operations on unsigned operands, including load, store, compare, test, arithmetic and logic operations. The branches that have significance for unsigned operand are listed in Table 2.15. Branch conditions not listed in this table are generally meaningless for signed operands and should not be used (for example, BPL).

The interpretation of the branch mnemonics after each type of operation listed in Table 2.15 is summarized in Table 2.16. Where a condition does not include the carry flag, unsigned overflow is not accounted for in the test. Thus, the actual condition could be wrong (for example, EQ is true but there was overflow, meaning the actual result may not be zero). When a condition includes C, it accounts for overflow, meaning the branch outcome reflects the true value of the result as if the overflow had never occurred. Load, store and test operations can't overflow.

Example 2.24.   Provide a sequence of instructions that will implement the operation

```
If X > Y

            Then Z=X-1

            Else Z=X+1
```

Table 2.15: Summary of branch conditions after unsigned data transfer, arithmetic, and compare instructions.

Branch Mnemonic	Condition	Complementary Mnemonic	Condition
EQ (equal)	Z=1	NE (not equal)	Z=0
HS (high or same)	C=0	LO (lower than)	C=1
HI (higher than)	C=0 and Z=0	LS (less than or same)	Z=1 or C=1

Table 2.16: Summary of branch conditions after unsigned data transfer, arithmetic, and compare instructions.

Branch Mnemonic	Meaning after Load/Store or Test of Operand	Meaning after Arithmetic operation	Meaning after a Compare of A to Operand (CMP)
EQ	Operand=0	Result=0	A=Operand
NE	Operand$\neq$0	Result$\neq$0	A$\neq$Operand
LO	No Meaning (carry	Overflow	A<Operand
HS	undefined)	No Overflow	A$\geq$Operand
HI	No Meaning (carry	Result>0	A>Operand
LS	undefined)	Result$\leq$0	A$\leq$Operand

assuming that X, Y, and Z are signed 8b memory operands at $0080, $00A1, and $00EE, respectively.

Solution: First, X and Y need to be compared to determine if X>Y. This can be accomplished with the CMP instruction, which performs the subtraction A-operand without saving the result.

```
LDA $80     ;load X into A
CMP $A1     ;compare it to Y
```

Next, a branch is needed. The comparison is of signed integers and is a greater than operation; thus, BGT is appropriate. The "Then" block will be the branch target; the "Else" will be the branch fall-through.

```
            BGT THEN
ELSE        INCA         ;X is still in A, increment it
            STA $EE      ;store result to Z
            BRA ENDIF    ;skip over Then to Else
THEN        DECA         ;X is still in A, decrement it
            STA $EE      ;store result to Z
ENDIF                    ;next program instruction here
```

A quick inspection reveals that both the Then and Else blocks end with the STA instruction. Thus, this instruction can be moved to the "ENDIF" line, where it will always get executed, saving two bytes of machine code, as shown below.

```
                LDA $80        ;load X into A
                CMP $A1        ;compare it to Y
                BGT THEN       ;If X > Y
        ELSE    INCA           ;Else increment X
                BRA ENDIF      ;
        THEN    DECA           ;Then decrement X
        ENDIF   STA $EE        ;store result to Z
```

Example 2.25.   Provide a sequence of instructions that will count the number of bits that are set in the byte at address $80. The count should be in accumulator A.

Solution: This is a repeat of Example 2.23. This time, however, a loop primitive instruction will be used. X will be used as a loop counter. A will be used to accumulate the number of 1's in the operand. The operand will be copied onto the stack to create a copy that can be modified without affecting the original value at $0080.

Answer:

```
            CLRA                ;initialize A to 0
            LDX $80             ;load operand into X
            PSHX                ; copy it to stack
            LDX #$8             ;initialize loop counter
    LOOP    LSL $01,SP          ;shift out a bit of operand ADC #0
                                ;accumulate it

            DBNZX LOOP ;loop 8 times
            PULX                ;remove scratch space from stack
```

2.11   LOOP PRIMITIVE INSTRUCTIONS

Loop primitive instructions combine a decrement, test and branch operation all in one instruction. They are used to create simple counter loops that count down from N to 1 (terminating when the counter reaches zero). They are related to branch and compare, but have been included in their own category because they also perform an arithmetic operation. The counter can be accumulator A, index register X or a direct or indexed addressed memory byte. The loop primitive instructions are summarized in table Table 2.17.

Table 2.17: Loop primitive instructions of the HCS08 CPU.

Instruction	Operation Type	Description
DBNZ opr, rel	Loop Primitive	Decrement operand; branch if result not equal to zero.
DBNZA rel	Loop Primitive	Decrement A; branch if result not equal to zero.
DBNZX rel	Loop Primitive	Decrement X; branch if result not zero.

2.12 MISCELLANEOUS INSTRUCTIONS

Table 2.18 lists the remaining HCS08 CPU instructions, some of which are covered in detail in subsequent chapters. They are listed here for completeness.

Table 2.18: Miscellaneous instructions of the HCS08 CPU.

Instruction	Operation Type	Description
BSR rel	Subroutine	Branch to a subroutine. Return address pushed onto stack.
STOP	CPU Control	Stops instruction sequencing and peripherals. See reference manual for complete description.
SWI	Interrupt	Software Interrupt.
JSR opr	Subroutine	Jumps to subroutine. Return address pushed onto stack.
RTI	Interrupt	Return from interrupt service routine.
RTS	Subroutine	Return from subroutine.
NOP	Miscellaneous	Do nothing (uses 1 cycle).
WAIT	CPU Control	Wait for interrupt.
BKGD	Debugging	

2.13 CHAPTER PROBLEMS

1. What are each of the HCS08 CPU registers (SP,PC,A,HX,CCR) used for?

2. After an ADD instruction executes, what does each of the following flag conditions indicate?

    a. V=1   b. C=0   c. H=0   d. N=0   e. Z=1

3. After subtracting signed operands with a SUB instruction, Z=1 and V=1. What is the value in accumulator A? What can be said about this result?

4. Accumulator A contains 44_{16}. The instruction SUB #$70 is executed by the CPU. What are the values in accumulator A and the flag registers after the instruction executes? (refer to the *HCS08 Family Reference Manual*).

5. For each of the following instructions, give the addressing mode, size of the operand and, where applicable, compute the effective address of the operand(s) (for each case assume HX contains 5555_{16})

   a. INCA

   b. LDA $AA

   c. SUB #$AA

   d. ASR $FFEE

   e. STHX $80

   f. ADD ,X

   g. MOV $FF,X+

   h. NSA

6. Describe in a sentence what each instruction in the previous problem does, using specific details of the instruction and describing all the changes made to registers and memory. Use the *HCS08 Family Reference Manual*. An example is "The instruction STA $80 copies the byte in accumulator A to memory location $0080; V is cleared and Z and N are set or cleared based on the value in A."

7. The 8 consecutive bytes in memory, starting at address $FEED, are $0A, $1B, $2C, $3D, $4E, $00, $87 and $FA. What are the effective address, size and value of the operand in each of the following load instructions; assume HX contains $FEF0.

   a. LDX $FEEF

   b. LDHX $FEED

   c. LDA ,X

   d. LDA $FFFE,X

   e. LDA $02,X

8. List 3 different ways to clear accumulator A with a single instruction.

9. List two different ways to toggle all of the bits in accumulator A with a single instruction.

10. Provide two sequences of instructions to compute 2x+1; assume X is an 8b unsigned value in memory at address $00F0.

11. ASR, LSR, ASL, LSL, ROR, ROL all operate on single bytes in memory. Suppose you needed to perform similar operations on 2 byte operands; for example, a logical shift left of the 16 bit value in memory starting at $0080 could be accomplished with

    `LSL $0081 ;shift lower byte left by 1, with msb out to carry`

    `ROL $0080 ;shift carry left into upper byte, with msb out to carry`

    This sequence will result in the most significant bit of the 16b operand in the carry flag, just as LSL would; however, the other flags might not be the same as the 8b LSL operation. For each 8b operation (ASR, LSR, ASL, LSL, ROR, and ROL) provide a similar sequence of two instructions to perform the equivalent 16b operation. Assume the operand is in memory at $0080. Determine if the condition of the flags N, Z, H, and V after each 16b operation (sequence) is equivalent to the condition of the flags after the equivalent 8b operation.

12. Repeat each sequence in the previous question, but assume the 16b operand is in memory at the address contained in HX.

13. Without using branch instructions, write a sequence of instructions to sign-extend the 8b operand at $0080 and store the 16b result to $0090 (Hint: SBC #0 will always result in $00 or $FF if A is $00).

14. The value in accumulator A can be represented as the 8b value (a_7 a_6 a_5 a_4 a_3 a_2 a_1 a_0). Write a sequence of instructions to make A contain (11 a'$_5$ 0 a_3 a'$_2$ a_1 0), where the notation a'$_5$ means the complement of bit a_5.

15. HX points to an array of 16b unsigned integers. Write a sequence of instructions to add 1 to the second integer (at HX+2).

16. HX points to an array of 16b unsigned integers. Write a sequence of instructions to add the first integer (at HX) to the second (at HX+2) and store the result to the third (at HX+4).

17. Repeat the last problem for subtraction of the first integer from the second.

18. Sketch a memory map showing the contents of the stack after the instruction sequence below. Assume HX contains $EEAB_{16}$, A contains 16_{16} and SP contains $00FF_{16}$. Label all memory addresses on the map and show the initial and final locations pointed at by SP (mark them $SP_{initial}$ and SP_{final}).

    PSHA

    PSHX

    PSHH

19. Write a sequence of instructions to place the decimal constants 32 and 17 on the top of the stack.

20. Suppose the top of the stack contains 4 bytes that are no longer needed. Write a single instruction to clean these bytes off the top of the stack.

21. Write a single instruction to create 3 new uninitialized bytes on top of the stack.

22. Suppose the top of the stack contains 3 unsigned 8b values. Write a sequence of instruction to compute the sum of these three bytes in accumulator A. At the end of the sequence, the three bytes should be removed from the top of the stack.

23. Using a loop, write a sequence of instructions to clear the 12 bytes in memory starting at the address contained in HX.

24. What does it mean if

    a. BGT is taken after the instruction CMP #$80?

    b. BHI is taken after the instruction CMP #$80?

    c. BEQ is taken after the instruction ADD #$80?

    d. BMI is taken after the instruction LDA $F0?

    e. BGT is taken after the instruction STA $F0?

    f. BLS is taken after the instruction TST $F0?

25. Write a sequence of instructions that does nothing (executes NOP) N times, where N is an 8b unsigned value at address $F000_{16}$, using a loop primitive instruction.

26. Repeat 25 without using a loop primitive instruction.

27. When performing division (shift right, for example), a CPU will truncate the answer. Write a sequence of instructions to compute Q/2, where Q is an 8b unsigned value in memory. The result, in accumulator A, should be rounded after the computation. (Hint: use ADC.)

28. What branches could you use to test for unsigned overflow?

CHAPTER 3

HCS08 Assembly Language Programming

Programming in assembly language is part skill and part art. Like any skill, assembly language programming improves with practice and experience. An experienced programmer can work more naturally and more quickly, ultimately to the extent that algorithms are conceived at the level of assembly language instructions. Most beginning assembly language programmers, however, start with an understanding of high-level programming language building blocks. Algorithms conceived at this level must be implemented in assembly language. This requires finding ways to translate high-level language building blocks in assembly language, often with consideration of efficiency in code size and execution time. In this chapter, assembly language programming is introduced from the perspective of translating or implementing algorithmic building blocks in assembly language.

This chapter is mostly about the mechanics of assembly language programming. This includes developing both the code, which implements the algorithmic structure of the program, and the data, which implements the data structures required. No attempt is made to cover algorithmic design or software design principles.

Assemblers differ in both the language syntax expected as well as the features provided; even assemblers for a single CPU architecture can differ significantly. Thus, it is necessary to study assembly language programming in the context of a specific assembler. In this book, the assembly language syntax of the Freescale HC(S)08/RS08 Macro Assembler is used.

3.1 ASSEMBLER STRUCTURE

An assembler is a program that translates assembly language input files into object code: machine code and data with additional information on program structure. Assemblers are simpler than compilers for high-level languages because much of the assembly process is a direct translation to machine code- there is generally a one-to-one correspondence between an assembly language instruction and a machine code instruction. Assemblers also provide the ability to use identifiers (or labels) in place of memory addresses to facilitate code development. Some assemblers also allow macros, which automate the inclusion of frequently used instruction sequences via textual substitution. More advanced features, such as code optimization, can also be provided. A disassembler is a program that converts machine code to assembly language. Because much of the structure of the assembly language input or object code is not available in the resulting machine code, complete and accurate disassembly is not always possible.

The input to an assembler is one or more text files containing assembly language. The primary output of the assembler is object code, which one can think of as a mixture of machine code and data with additional information which can include information on the structure of the code and data areas of the program, symbolic information about variables and subroutines and other information as needed. For embedded systems, object code generally consists of sequences of bytes of machine code and data along with information about where the bytes are to be located in memory. The operating system on a general-purpose computer requires that object code be converted into an executable file format before it can be run. This executable file format can include information to describe the environment in which the program is to be run, additional operating system services it requires, debugging information and other information needed to prepare the program to be run. This additional information allows the operating system to perform automated loading and execution of the program, enabling multiple programs to exist simultaneously in memory without having to plan beforehand where they will be located. Small embedded systems, in contrast, generally execute a specific set of programs and often lack operating system support. Programs are explicitly placed at specific memory locations; therefore, the object file format often describes only enough information to determine where the program is to be placed in memory. Other required services or facilities, normally provided by an operating system on a general-purpose computer, are explicitly handled by the programmer.

Assemblers also produce human-readable output that can be useful for debugging and analysis of the resulting machine code. One common assembler output file is a listing file, which shows how the assembler interpreted the assembly language. The listing file shows each line of the assembly language input along with the memory addresses resolved by the assembler, resulting machine code or data and other diagnostic information. As assembly language is a human-readable form of machine code, a listing file is like a human-readable form of object code.

3.2   MACHINE LANGUAGE ENCODING

There is direct correspondence between HCS08 CPU assembly language mnemonics and their corresponding opcodes. Table 3.1 shows the mapping between different forms of the LDA instruction and the corresponding machine code, using the notation described in the *HCS08 Family Reference Manual*. Opcodes are 1 or 2 bytes and are shown in upper case notation; the part of the machine code specifying the data needed to encode an operand is shown in lowercase. Each letter (uppercase or lowercase) corresponds to one hex digit in the machine code. Once can count the number of pairs of hex digits to determine the number of bytes of machine code each instruction requires.

Each HCS08 CPU opcode specifies both the operation to be performed as well as the addressing mode(s) for the operands. The opcode byte $A6_{16}$, for example, specifies the instruction "load immediate operand into A" while the opcode byte $B6_{16}$ specifies the instruction "load direct operand into A." For each addressing mode, the opcode byte(s) are followed by the data needed by the addressing mode. The meaning of the letters is described in the *HCS08 Family Reference Manual*.

Table 3.1: Machine Code Mapping for Different Operand Formats for LDA Instruction.

Instruction Form	Addressing Mode	Machine Code	
LDA #opr8i	IMM (immediate)	A6	ii
LDA opr8a	DIR (direct)	B6	dd
LDA opr16a	EXT (extended direct)	C6	hh ll
LDA oprx16,X	IX2 (16b offset indexed)	D6	ee ff
LDA oprx8,X	IX1 (8b offset indexed)	E6	ff
LDA ,X	IX (indexed)	F6	
LDA oprx16,SP	SP2 (16b SP offset)	9ED6	ee ff
LDA oprx8,SP	SP1 (8b SP offset)	9EE6	ff

For example, the byte *ii* indicates an 8b immediate, *dd* an 8b direct address and *ee* the MSB of a 16b offset.

Table 3.2 shows the machine language encoding for 4 of the branch instructions. The opcode encodes the type of branch (branch condition). The 8 bit offset (*rr*) needed for the relative addressing follows the opcode.

Table 3.2: Machine Code for a Subset of the Branch Instructions.

Instruction Form	Adressing Mode	Machine Code	
BCS rel	Relative	25	rr
BEQ rel	Relative	27	rr
BGE rel	Relative	90	rr
BGT rel	Relative	92	rr

Example 3.1. What is the machine code for
    a) LDA #$32   b) LDA $32   c) LDA 128,X
Solution:
a) This is the immediate form of the load. The opcode is thus $A6_{16}$ and it is followed by the immediate operand (ii), which is 32_{16}. Thus, the machine code is $A632_{16}$.
b) This is the direct form of the load. The opcode is $B6_{16}$. This is followed by the 8b direct address (dd), which is 32_{16}. Thus, the machine code is $B632_{16}$.

c) This is an offset indexed form. Since the offset (128_{10}) is small enough to be encoded as an unsigned 8b integer, 8b offset indexed is chosen. The opcode, $E6_{16}$ is followed by the 8b offset 80_{16}. Thus, the machine code is $E680_{16}$.

Answer: a) $A632_{16}$ b) $B632_{16}$ c) $E680_{16}$

Example 3.2. The instruction BRA $EE00 is to be assembled and loaded in memory starting at address $EE0C_{16}$. What is the machine code?

Solution: The opcode for BRA is 20_{16}. The effective address is computed as

$$\text{Target Address} = PC_{branch} + 2 + \text{offset}.$$

Rearranging, we get

$$\text{offset} = \text{Target Address} - PC_{branch} - 2.$$

Thus, offset=$EE00_{16} - EE0C_{16} - 0002_{16} = FFF2_{16}$. The machine code offset is $F2_{16}$ (the upper byte $FF consist of sign bits and can be ignored).

Answer: $20F2_{16}$

Example 3.3. The machine code $92F3_{16}$ is a branch in memory starting at address $F004_{16}$. What is the address of the branch target?

Solution: The effective address is computed as $PC_{branch} + 2 + \text{offset}$, where the offset is signextended to 16b by the CPU at run time. $F004_{16} + 0002_{16} + FFF3_{16} = EFF9_{16}$. Thus, the branch target address is $EFF9_{16}$.

Answer: $EFF9_{16}$

Example 3.4. The instruction BRCLR 3,$AB,$EE62 is to be assembled and located in memory starting at address $EE0C_{16}$. What is the machine code?

Solution: The opcode for BRCLR depends on which bit is to be tested. For bit 3, the opcode is 07_{16}. The opcode is followed by the direct address byte and then by the relative byte. The direct address byte is AB_{16}. Because there are 3 bytes of machine code for the BRCLR, the effective address is computed as

$$\text{Target Address} = PC_{branch} + 3 + \text{offset}.$$

Thus, offset=$EE62_{16} - EE0C_{16} - 0003_{16} = EE62_{16} - EE0F_{16} = FF53_{16}$. Thus, the offset is 53_{16}.

Answer: $07AB53_{16}$

3.2.1 ASSEMBLY LANGUAGE FILE FORMAT

The HC(S)08/RS08 Macro Assembler file format, like that of many assemblers, is line-oriented; each line in the file represents one assembly language statement. Assembly language statements cannot span multiple lines. The main components of a line are label, opcode, operand(s) and comment; thus, an assembly language statement (line) has the format

```
[label[:]]    [operation [operand(s)]] [;comment]
```

All of the elements on the line are optional but, if present, must be in the order specified. They can be present in different combinations depending on the operation being specified. The exception is that operand(s) are not allowed unless an operation is specified. Whitespace between the fields is allowed, but not within (for example, not between operands).

The optional label, if present, must start in the first column and is case insensitive. The label corresponds to the memory location at which the machine code (or data bytes) generated from the assembly language line will start. In order for the assembler to distinguish labels from operations, the syntax requires that operations never start in the first column; that is, any identifier that begins in the first column is interpreted as a label. Therefore, lines without labels must be indented. The label is separated from the operation by whitespace or a colon, neither of which are part of the label. Labels must start with a letter, can be at most 16 characters in length and can contain letters, numbers, dash ('-') or underscore ('_').

The operation, if present, may be followed by operands. Operations are assembly language mnemonics, assembler directives or assembler pseudo-ops. A directive is a command to direct or change the assembler processing, such as to include other files. A pseudo-operation (pseudo-op) is like an assembly language mnemonic, but it does not generate machine code. Pseudo-ops are used to organize program and data areas.

Operand types and syntax depend on the operation(opcode or pseudo-op) specified. Where constants are allowed in operands, simple constant expressions can be used. HC(S)08/RS08 Macro Assembler expressions use ANSI C type operators and precedence; some of the operators supported are shown in Table 3.3. These constant expressions are evaluated by the assembler and, since they evaluate to a constant, do not generate any additional machine code compared to the programmer performing the computations manually.

A subset of the HC(S)08/RS08 Macro Assembler pseudo-ops are listed in Table 3.4. The ORG pseudo-op defines the memory address that the assembler should assign to the object code byte generated (machine code or data). The assembler will start assigning subsequent bytes of object code to consecutive memory locations starting from this address, until it processes another ORG pseudo-op. If no ORG is given, the assembler begins placing object code at address 0000_{16}.

EQU is used to define program constants, which can improve code readability and facilitate code maintenance. Repeatedly used constants should be assigned a name and then, if the constant value needs to be changed, only a single modification to the assembly language program needs to be made (by changing the EQU pseudo-operation). Each time the assembler finds the label in the

Table 3.3: Operators that can be used in constant assembly language expressions.

Operator	Operation	Examples
*, /	Multiplication, Division	2*RAM_SIZE, PI/3
+, -	Addition, Subtraction	RAM_START+RAM_SIZE - 1
-	Negation (unary)	- CONSTANT
<<	Logical Left Shift by n	RAM_SIZE<<2
>>	Logical Right Shift by n	RAM_SIZE>>3
%	Remainder After Division	QUANTITY%2
&	Bitwise AND	PORT&MASK
|	Bitwise OR	MASKA|MASKB
^	Bitwise XOR	DATA^$FF
~	Ones Complement	~MASK
!	Bitwise Logical NOT	!MASK

assembly language file, it replaces the label with the associated constant value. Examples of usage of EQU are shown in Code Listing 3.1.

```
1   RAM_START    EQU    $0080       ; define starting address of RAM
2   ALARM_MASK   EQU    %00010000   ; define bit to control alarm
3   PI           EQU    %11001001   ; PI, to 6 binary places
```

Code Listing 3.1: Equate Pseudo-Op Examples.

Given these definitions, one could write, for example, LDA #PI instead of LDA #%11001001; the former instruction is much clearer and provides more information to the reader about the intent of the programmer and the function of the code (%11001001 is the 8b fixed-point value of pi to 4 binary places). Similarly, one could write ORG RAM_START instead of ORG $0080, which again provides additional information about the intent of the programmer to begin the next section at the start of RAM; without this, someone reading the assembly language file would need to know that $0080 is the expected start of RAM, which is not true for every HCS08-based microcomputer.

The storage pseudo-ops DS, DC.B, and DC.W are covered in more detail in Section 3.4.

3.2.2   ASSEMBLER PROCESSING

Understanding the processing that the assembler carries out on an assembly language file is not required to program in assembly language; however, having some familiarity with the process can help the programmer to understand the limitations of the assembler as well as to diagnose errors when they arise.

Pseudo-Op	Format	Function	Example
ORG	ORG *address*	Sets the origin (beginning address) of the next assembly language line to address.	ORG $EE00
EQU	label: EQU n	Sets the value of "label" to be equal to n. 'n' can be a number, expression, or another label. Used to assign pseudonyms to program constants.	msb: EQU $80
DS	DS *count*	Define space. Reserves *count* bytes of uninitialized space. Used for creating uninitialized data areas and to align memory areas.	DS 5
DC.B	DC.B *list*	Define constant byte(s). Reserves a single byte of storage for each element of *list*, initialized to the value of the element. A comma separated list of constants can be provided.	DC.B 17,'a',$45
DC.W	DC.W *list*	Define constant word(s). Reserves a word (two bytes) of storage for each element of *list*, initialized to the value of the element. A comma separated list of constants can be provided.	DC.W $FFFE,2
DCB.B	DCB.s *c,v*	Define constant block. *s* (size) is either B or W (byte or word, respectively). *c* (count) is number of elements reserved. *v* (value) is the initialization value. For initializing blocks of memory to a single value.	DCB.B 32,0

Table 3.4: HC(S)08/RS08 Macro Assembler Pseudo-Operations.

To generate object code, an assembler translates assembly language instructions into sequences of machine code bytes, data pseudo-ops into sequences of data bytes, and determines the address of each. A major part of this process involves resolving labels into addresses. Since most assembly language statements translate into a fixed number of bytes, it is relatively straightforward for the assembler to incrementally generate address information.

Consider, for example, the assembler input shown in Code Listing 3.2 (note that line numbers are included for reference and are not part of the program). The assembler reads line 1 and associates the identifier PROG_AREA with the constant $EE00_{16}$; subsequently, each time PROG_AREA appears

```
1    PROG_AREA         EQU   $EE00          ; location of code in memory
2
3                      ORG   PROG_AREA
4    START             LDX   #10
5    NEXT              LDA   $00
6                      BNE   NEXT
```

Code Listing 3.2: Simple Assembly Language Example Program.

in the file the assembler replaces it with $EE00_{16}$. The assembler maintains a symbol table to keep track of all such associations. Line 2 contains only whitespace and is skipped. Line 3 is an origin pseudo-op with the operand PROG_AREA; this line is therefore equivalent to ORG $EE00, so the assembler resets its *location counter* to $EE00_{16}$. The location counter is simply a variable used by the assembler to keep track of the current object code address.

Line 4 is processed next. This is the first assembly language line to actually generate object code. The label START corresponds to the current value of the location counter, which is still $EE00_{16}$; thus, the assembler adds this to its symbol table. The instruction LDX #10 translates directly to the two bytes of machine code $AE0A_{16}$, and these two bytes become the first two bytes of object code. The location counter is incremented by two, becoming $EE02_{16}$. The assembler then processes line 5. NEXT is added to the symbol table and is equivalent to the current value of the location counter ($EE02_{16}$). Two bytes of object code are generated for LDA $00 (corresponding to the machine code $B600_{16}$), and the location counter is updated to $EE04_{16}$. Finally, line 6 is processed. The opcode for the BNE instruction is 26_{16}. From its symbol table, the assembler associates the branch target NEXT with address $EE02_{16}$. The relative offset is computed as $EE02_{16}-EE04_{16}-2=FC_{16}$. Thus, two bytes of object code $26FC_{16}$ are generated and the location counter is updated to $EE06_{16}$. At this point, the assembler encounters the end of the input file and writes its output files.

In the above example, it is possible to determine the exact object code bytes for each instruction as well as the address as instructions are processed. In some instances, however, it is not possible to do this. Consider, for example, when the assembler encounters a forward branch, as in line 5 of the program in Code Listing 3.3. Because label POSITIVE is not yet in the symbol table (it is not encountered until line 7), the assembler cannot determine the exact machine code for the BGE instruction. However, it is known that the BGE POSITIVE instruction has two bytes of machine code, so the assembler can continue processing with the branch target unknown, and return later to complete the computation of the relative offset once the address of the target is in the symbol table. Such instances require the assembler to make another pass over the program.

A slightly more difficult situation occurs when the assembler cannot determine the size of the machine code because of forward references. The only instance in the HCS08 when the assembler cannot determine the exact size of the machine code without knowing the value of a symbol is when either direct or extend addressing is called for. Consider the simple program in Code Listing 3.4.

```
1   PROG_AREA      EQU     $EE00          ; location of code in memory
2
3                  ORG     PROG_AREA
4   START          LDA     $00
5                  BGE     POSITIVE
6   NEGATIVE       NEGA
7   POSITIVE       ADD     #14
```

Code Listing 3.3: Assembly Language Program with Forward References.

```
1                  ORG     $00F7
2   START          LDA     $00
3                  BGE     POSITIVE
4                  JMP     NEGATIVE
5   POSITIVE       ADD     #14
6   NEGATIVE       SUB     #14
```

Code Listing 3.4: Assembly Program Requiring Speculative Assembly.

The instruction JMP NEGATIVE, refers to the label NEGATIVE, which has not been encountered yet. Until the assembler knows the address associated with NEGATIVE, it cannot determine if direct or extended addressing can be used (i.e., whether or not the address is in the zero page). Because extended addressing requires two bytes to encode the address (compared to 1 byte for direct), the size of the object code cannot be determined. On the other hand, the assembler needs to know the size of the instruction in order to update the location counter and ultimately determine the address to associate with NEGATIVE. In the program in Code Listing 3.4, an iterative assembly may never terminate because of the mutual dependence of the address of label NEGATIVE and the size of the machine code JMP NEGATIVE. In such cases, a method of breaking the infinite assembly loop is required, the easiest of which is to simply use extended addressing whenever the forward reference cannot be easily resolved. This is apparently the method used by HC(S)08/RS08 Macro Assembler, which assembles the code using extended addressing for the JMP instruction. It should be noted, however, that in most instances it is possible for an assembler to correctly resolve forward references that involve direct addressing to data areas.

3.2.3   LISTING FILE

An example of a listing file output by HC(S)08/RS08 Macro Assembler is shown in Code Listing 3.5. Recall that the listing file is a diagnostic output file that provides the programmer with information on how the assembly language input was processed by the assembler; it is, in a sense, a mapping between the object code and the assembly language input. The operation of the program will not be

```
1    Freescale HCS08-Assembler
2    (c) Copyright Freescale 1987-2006
3
4    Abs. Loc  Obj. code     Source line
5    ----  --------------   -----------
6        1      0000 0080    DATA_AREA        EQU      $0080
7        2      0000 EE00    CODE_AREA        EQU      $EE00
8        3      0000 0000    PORTA            EQU      $00
9        4      0000 0024    MASK             EQU      %00100100
10       5
11       6                                    ORG      CODE_AREA
12       7  EE00 3F 80       START            CLR      COUNT
13       8  EE02 B6 00       MAINLOOP         LDA      PORTA
14       9  EE04 A4 24                        AND      #MASK
15      10  EE06 26 02                        BNE      SKIP
16      11
17      12
18      13  EE08 3C 80                        INC      COUNT
19      14  EE0A 12 01       SKIP             BSET     1,$01
20      15  EE0C 13 01                        BCLR     1,$01
21      16  EE0E 20 F2                        BRA      MAINLOOP
22      17
23      18                                    ORG      DATA_AREA
24      19  0080             COUNT            DS       1
25      20  0081 11                           DC.B     17
```

Code Listing 3.5: List File Example.

considered at this time, only how it is interpreted by the assembler. Once again, line numbers have been included for reference but are not part of the actual file. There is one line in the listing file for each line of assembly language input (lines 6-25 in Code Listing 3.5), plus some additional lines added by the assembler (lines numbered 1-5). Each listing file line begins with the location counter value for that line, which is equivalent to the address that any object code produced from that line will be placed. Some lines have the same location counter value because they assemble into zero bytes of object code. Notice that until an origin pseudo-op appears in the assembly language input file, the assembler location counter is incremented from 0000_{16}. For assembly language instructions, the line includes a number in square brackets, corresponding to the number of clock cycles the CPU takes to execute the instruction. The third column represents the object code produced (object code has been shown in bold in the code listing). For assembly language instructions, this is the machine code of the instructions; for initialized data areas (line 21), the object code bytes are the data bytes. Finally, the assembly language input line itself is listed.

3.2.4 OBJECT FILE: S19 FILE FORMAT

Like many assemblers for Freescale processors, the HC(S)08/RS08 Macro Assembler can generate *S19 file format* object files. The S19 format is a very simple line-oriented format that lists multiple sequential object code bytes on each line with the starting address of the line being given. Each line is called an S-Record. The S19 file corresponding to Code Listing 3.5 is provided in Code Listing 3.6. The spaces in each line have been added to simplify the presentation and are not part of the S19 file itself. There are five fields per line, some of which may not be present. These are TYPE, LENGTH, ADDRESS, CODE and CHECKSUM. There are three types of records common for 8 bit processors. A block of S-Records begins with an S0 (header) record and ends in an S9 (termination) record. Neither the S0 nor S9 records contain any object code. S0 records allows a description or header to be added to a block of S-Records. The S1 records contain the actual bytes of object code.

```
1   S1   13   EE00   3F 80 B6 00 A4 24 26 02 3C 80 12 01 13 01 20 F2   A4
2   S1   05   0080   00 11   69
3   S9   03 00 00   FC
```

Code Listing 3.6: S19 Record Listing.

The second field in an S-Record is the length field, which describes the number of bytes listed in the record, including address, code, data and checksum bytes. For HCS08 processors, addresses are 2 bytes. Following the address is the object code. Because addresses are two bytes and checksum is 1 byte, there are always 3 fewer bytes of object code in the S-Record than the length specifies. The object code represents sequential bytes in memory, starting at the address listed in the address field. In the first S1 record in Code Listing 3.6, for example, the byte $3F_{16}$ is to be placed at address $EE00_{16}$, 80_{16} at $EE01_{16}$, and so on. The final byte of the line is the one's complement of the 8 bit sum of the bytes on the line. This can be used for error checking when the record is transmitted over a serial link, where link errors may be introduced.

The memory maps that result from the object code described by the S19 record file in Code Listing 3.6 are shown in Figure 3.1. Memory locations not shown or shown as empty are not affected by this S19 file.

3.3 ORGANIZING PROGRAM AND DATA AREAS

In an embedded system, the microcomputer is typically running programs directly from programmable ROM memory (Flash or EEPROM), while RAM is used for program variables and the

EE00	3F	0080	00
EE01	80	0081	11
EE02	B6		
EE02	00		
EE03	A4		
EE04	24		
EE05	26		
EE06	02		
EE07	3C		
EE08	80		
EE09	12		
EE0A	01		
EE0B	13		
EE0C	01		
EE0D	20		
EE0E	F2		

Figure 3.1: Detailed memory specified by Code Listing 3.6.

system stack. In some instances, however, RAM might be used for both programs and data, especially if RAM is abundant and the programs are loaded from secondary storage or over a communications link. The exact organization of programs and data in an embedded system depends on the resources available, application requirements and programmer preference. However, two common program organizations are shown Figure 3.2. The first organization is for systems that use RAM for combined data and programs. In this organization, programs are generally placed starting at the bottom of

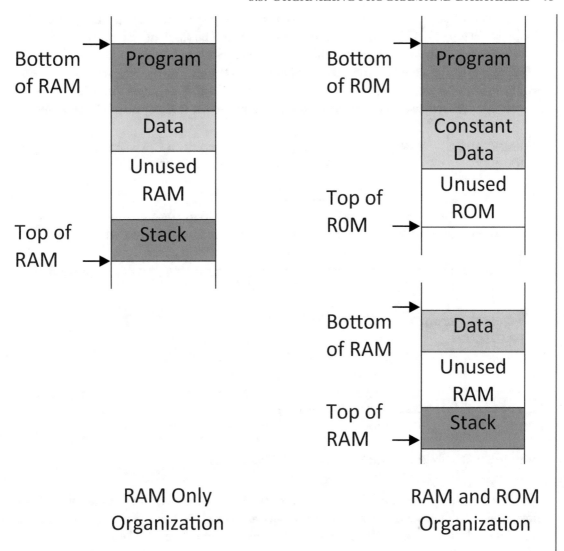

Figure 3.2: High level memory maps of program, data and stack areas.

RAM, followed by program data. The stack begins at the top of RAM and grows downward into the unused RAM space between the top of the stack and the end of the data (though nothing prevents a runaway program stack from growing further). The second organization shows how a program might be organized in a combined RAM and ROM system. This is a more typical organization. The program code and constant data could be stored at the bottom of the ROM memory region. RAM is used in a similar way to the first organization without the code at the bottom.

In either of the two program organizations shown in Figure 3.2, the stack pointer is initialized to point to the top of RAM at the beginning of the program. The absolute location of the data area is usually not necessary to know because the assembler uses labels to refer to specific data locations. Therefore, the data area can be started at the lowest free RAM area. The assembly language skeleton program in Code Listing 3.7 implements the RAM-only program organization shown in Figure 3.2. The first two lines define the parameters for the RAM area in the memory map. The ORG pseudo-

```
1   RAMSTART        EQU       $0060       ;Address of Start of RAM
2   RAMSIZE         EQU       $0100       ;Size of RAM, in Bytes
3
4                   ORG       RAMSTART
5   ;-----------------------------------------------------------
6   ;The following code initializes SP to the top of RAM
7   ;note: TXS subtracts 1 from HX before copying into SP
8   ;-----------------------------------------------------------
9   MAIN:           LDHX      #{RAMSTART+RAMSIZE}
10                  TXS
11  ;Insert Program Code Below
12  MAINLOOP:       NOP                     ;do nothing
13                  BRA       MAINLOOP      ;forever
14
15  ;-----------------------------------------------------------
16  ;Define Program Data Variables Below
17  ;-----------------------------------------------------------
```

Code Listing 3.7: Skeleton Assembly File for RAM Only Organization.

op on line 4 is the only ORG required for this program organization because the data area begins immediately after the code area. The code on lines 12 and 13 would be replaced by the actual program code. Data variables would be defined, with labels, beginning after line 17. The stack computation on lines 9 and 10 initializes the stack pointer to point to the last valid RAM address. The last RAM address is computed as RAMSTART + RAMSIZE −1; the assembler adds the two constants in the operand expression on line 9; and the TXS implements the -1 needed in the expression. Notice how the assembler has been called upon on line 9 to compute the last address in RAM; this constant expression does not introduce any additional machine code instructions or bytes; the LDHX #{RAMSTART+RAMSIZE} instruction is equivalent to LDHX #$0160 once the assembler computes the constant expression. However, if the program is ported to another HCS08-based microcomputer with a different memory organization, only the two equates on lines 1 and 2 would need to me modified. The stack should be initialized whether or not it is used by the program because the CPU also uses the stack for interrupts.

A skeleton program to implement the combined RAM and ROM memory program organization is shown in Code Listing 3.8. This program is similar to the previous one except for the two ORG pseudo-ops, which are required to locate the separate code and data areas into ROM and RAM, respectively.

```
1   RAMSTART        EQU         $0060    ;Address of Start of RAM
2   RAMSIZE         EQU         $0100    ;Size of RAM, in Bytes
3   ROMSTART        EQU         $F000    ;Start address of ROM
4
5
6                   ORG         ROMSTART
7   ;------------------------------------------------------------
8   ;The following code initializes SP to the top of RAM
9   ;note: TXS subtracts 1 from HX before copying into SP
10  ;------------------------------------------------------------
11  MAIN:           LDHX        #{RAMSTART+RAMSIZE}
12                  TXS
13  ;Insert Program Code Below
14  MAINLOOP:       NOP                         ;do nothing
15                  BRA         MAINLOOP    ;forever
16  ;------------------------------------------------------------
17
18                  ORG         RAMSTART
19  ;------------------------------------------------------------
20  ;Define Program Data Variables Below
21  ;------------------------------------------------------------
```

Code Listing 3.8: Skeleton Assembly File for RAM and ROM Organization.

3.4   CREATING AND ACCESSING VARIABLES

A *variable* can be defined as one or more consecutive memory locations assigned to hold a data value of a particular size and type. Since the memory locations are consecutive and assigned to a single variable name, the variable can then be uniquely specified by its address (the address of the first byte in the sequence). In assembly language, an identifier (label) can be associated with this address to assign the variable a name. Implementation of algorithms in assembly language is as much about defining, organizing and accessing data as it is about writing assembly language instruction sequences. In assembly language, the programmer does not define the type of a variable (integer, character, etc.) only its size; how the program manipulates the variable defines its type. Many embedded systems have limited memory available for programs and data and its careful use prevents excessive memory consumption and leads to more efficient programs. This is one advantage assembly language has over high-level languages.

To create variables in assembly language it is necessary to reserve the required amount of memory for each variable in the data area of the program. For uninitialized variables, the define storage (DS) pseudo-op reserves the required number of bytes. By including a label with each DS pseudo-op, the programmer can refer to the variable's address using that label. As a result, the programmer does not need to know exactly which address will be assigned to the variable and, if the program is later modified, does not need to update every reference to the variable in the program.

Examples of using DS to create uninitialized variables are shown in Code Listing 3.9. The ORG

```
1                    ORG     $0080
2    SPEED           DS 1    ;an 8b variable
3    TEMPERATURE     DS 2    ;a 16b variable
4    VOLTAGE         DS 4    ;a 32b variable
```

Code Listing 3.9: Assembly Example-Defining Variables.

statement defines the starting address of the data; SPEED is defined as an 8b variable, TEMPERATURE a 16b variable and VOLTAGE a 32b variable. All storage must be allocated in multiples of bytes, although a programmer is free to manipulate the individual bits of these bytes as separate variables. The DS pseudo-ops are used to reserve bytes so that addresses can be correctly assigned to the labels; these addresses, in turn, appear as extended, direct or immediate (pointer) operands in the machine code. There is, however, no object code generated. The above storage definitions result in the memory map in Figure 3.3, where the address of SPEED is $0080, the address of TEMPERATURE is $0081 and the address of VOLTAGE is $0083. Since the storage is uninitialized, however, no values need to be loaded into memory; thus, variables defined using DS create *zero bytes of object code*. When the program is executed, whatever bytes happen to be in the memory locations occupied by these variables define their uninitialized state. It is the programmer's responsibility to properly initialize all variables.

Programming languages define different categories of variables depending on how they are shared among program entities (subroutines). Global variables are variables that are not limited to any particular subroutine and are equally accessible by all parts of the program. Global variables can be defined in the data area of a program since there is a single copy of each and the amount of space required can be determined at the time of assembly. Local variables, on the other hand, belong to the subroutine in which they are defined and are only needed during the subroutine execution (they do not retain their value across subroutine calls). Because not all subroutines are active at any given time, it would be wasteful to assign fixed memory locations to local variables for the duration of the program. Therefore, local variables are usually created on the stack. These stack variables are covered in Section 3.6.2.

$0080	XX	SPEED
$0081	XX	TEMPERATURE
$0082	XX	
$0083	XX	VOLTAGE
$0084	XX	
$0085	XX	
$0086	XX	

Figure 3.3: Detailed memory map of variables defined in the Code Listing 3.9.

There are other qualifiers that some languages place on variables. These include the qualifiers static and constant in front of variable names. Though there are various definitions of a static variable, in C a variable declared as static holds its value across subroutine calls; thus, static variables provide state for subroutines. No matter how many times a subroutine is nested, there is only one copy of a static variable. Therefore, a static variable in assembly language can be implemented in the same way as global variable.

A constant variable is a variable whose value does not change over the course of a program. The assembly language programmer can define constants in two ways: with equate pseudo-ops (EQU), accessing them using immediate addressing, or with a separate constant data area within the code or data areas of the program (since the values are constant, they can be stored with the code in ROM). Technically, the former method is not creating a variable since the constant is not kept in a single memory location that can be referenced by address.

Methods for defining and accessing constant and global program variables are shown in Code Listing 3.10, which demonstrates a program to compute Y=aX+b using 8b unsigned arithmetic to produce a 16b result. Because these constant and variable names might conflict with register names, the variables X and Y are called vX and vY; similarly, the constants A and B are called cA and cB. Variables vX and vY are declared as uninitialized global variables on lines 37 and 38. Because the arithmetic can produce a 16b sum, vY is assigned 2 bytes of memory. Constant cB is declared on line 14 using an EQU pseudo-op. In contrast, constant cA is declared as an initialized constant variable at the end of the program area using a DC.B pseudo-op. The program begins by initializing the stack on lines 17 and 18, even though the stack is not directly by the program. Line 19 illustrates how the global variable vX is accessed by its label; since vX and vY are declared in the memory region starting at $0080, direct addressing is used by the assembler to access them. Constants declared as variables, like cA in the program, are accessed in the same way as variables (using direct or extended

```
1  ;-----------------------------------------------------------
2  ;Sample Program to Illustrate Use of Constant Variables
3  ;
4  ;program computes vY=cA*vX+cB, where
5  ;         cA and cB are 8b unsigned program constants
6  ;         vX and vY are unsigned 8b and 16b global variables
7  ;
8  ;The value of X must be initialized outside of the program
9  ;overflow is ignored
10 ;-----------------------------------------------------------
11 RAMSTART EQU  $0060   ;Address of Start of RAM
12 RAMSIZE  EQU  $0100   ;Size of RAM, in Bytes
13 ROMSTART EQU  $F000   ;Start address of ROM
14 cB       EQU  15 ;Define constant cB
15
16          ORG  ROMSTART
17          LDHX       #{RAMSTART+RAMSIZE}
18          TXS        ;initialize SP
19 START:   LDA  vX    ;load vX into Acc.
20          LDX  cA    ;load A into X
21          MUL        ;multiply X:A=X*A
22          ADD  #cB   ;add 8b constant to LSB
23          STA  vY+1  ;store LSB of result
24          TXA        ;move result MSB into Acc
25          ADC  #0    ;add carry from LSB
26          STA  vY    ;store to msb
27 STOP     BRA  STOP  ;idle here forever
28 ;-----------------------------------------------------------
29 ;Program Constant Variables
30 ;-----------------------------------------------------------
31 cA       DC.B $32
32
33          ORG  RAM_START
34 ;-----------------------------------------------------------
35 ;Program Global Data Variables
36 ;-----------------------------------------------------------
```

Code Listing 3.10: Sample Program Illustrating the Use of Constant Variables.

addressing). In this case, cA has been declared in the program area, which is not in the zero page, so extended addressing is used. Notice the difference in how cB, which is declared using an EQU pseudo-op, is accessed on line 22. Since cB is truly a constant (not a constant variable) it must be accessed using the immediate notation #cB.

In the program, the 16b product of vX and cA is in X:A. The constant cB is added to this sum using extended arithmetic; cB is added to the LSB that is in accumulator A (line 22) and the carry out of this sum is added to the MSB (line 25) using and ADC instruction. Note that the two intervening instructions do not affect the carry flag, so the carry is preserved across the addition. Since vY is 2 bytes, the MSB is at address vY while the LSB is at address vY+1. Thus, when accessing the LSB in line 23, the assembler expression vY+1 is used to refer to its address. The assembler computes the address and encodes it as a direct operand ($82).

The object code generated for this program, in hex dump format, is shown in Code Listing 3.11, where the program constants and the instructions accessing them have been highlighted with bold font. The first three lines consist entirely of machine code for the program. The con-

EE00: 45 01 00 94 C6 00 80 **CE**

EE08: **EE 18** 42 **AB 0F** C7 00 82

EE10: 9F A9 00 C7 00 81 20 FE

EE18: **32**

Code Listing 3.11: Memory Dump Format of Machine Code Generated for Code Listing 3.10.

stant cB, defined to the value $0F_{16}$ using an EQU pseudo-op, appears in the machine code of the ADD instruction (**AB0F**). The constant variable cA, which is stored in a separate area of memory at the end of the program, generates one byte of object code at EE18 whose value is the value of the constant 32_{16}. This constant variable is referenced by the machine code instruction $\text{C}\textbf{EE18}_{16}$ (LDX cA) using extended addressing.

Because constant variables may be accessed using extended addressing, they can require up to two bytes to encode into machine code operands. Constants defined as 8b constant variables require one byte to encode as immediate operands. It usually requires an extra clock cycle to access an extended operand compared to an immediate operand, due to the need to load the operand from memory (recall that the immediate is in the machine code). Thus, for the most part, 8b constants are more efficiently implemented with EQU pseudo-ops, considering both object code size and execution time. For 16b constants, the size of the machine code is equal in both cases, but the access time for extended addressing is still larger. There are no immediate operands larger the 16b in the HCS08, so any constant greater than 16b must be defined as a constant memory variables.

Example 3.5.   What are the object code bytes generated by the following program?

```
1                     ORG   $EE00
2     START     LDA   VARIABLE
3                     ADD   #13
4                     STA   VARIABLE
5                     BRA   START
6                     ORG   $0080
7     VARIABLE  DS    2
```

Solution: Use the *HCS08 Family Reference Manual*. Scanning the code, it can be determined that VARIABLE and START are the only labels. The address corresponding to START is clearly $EE00_{16}$ and the address of VARIABLE is 0080_{16}. LDA VARIABLE can be encoded with direct addressing; its machine code is $B680_{16}$. ADD #13 uses immediate addressing; its machine code is $A90D_{16}$. STA VARIABLE is also direct addressing; its machine code is $B780_{16}$. Finally, BRA START uses relative addressing; to compute the address, we need to find the object code address of BRA, which can be determined by counting the number of machine code bytes from the start of the code: $EE00_{16}+2+2+2=EE06_{16}$. The relative offset is thus $START-EE06_{16}-2$ or $F8. The BRA START machine code is thus $20F8_{16}$. The DS pseudo-op produces no object code.

Answer: The object code bytes starting from address $EE00_{16}$, are (in hex) B6, 80, A9, 0D, B7, 80, 20, and F8.

3.4.1   TABLES, ARRAYS AND DATA STRUCTURES

In assembly language, only the size of an operand needs to be specified to the assembler. Because tables and arrays are made up of more than one similar elements, they are equivalent structures in assembly language. A data structure is similar to a table in that it represents a collection of elements; it differs in that these elements can have different sizes and types.

To create an initialized table (or array), the DC.B or DC.W directives are used (depending on whether the elements are 8b or 16b). A comma-separated list of the initial values of the table elements is then provided, with the length of the list determining the size of the array. The type of elements (signed or unsigned) as well as the length of the array are not explicitly declared. There is nothing to stop a program from reading or writing past the end of the array except good programming. For example, to declare an array called ODDS of ten 8b values initialized to the positive odd integers one would write

```
ODDS:       DC.B 01,03,05,07,09,11,13,15,17,19
```

The total size of this array is 10B. To define 16b values, one would use DC.W, as in

```
EVENS:      DC.W 02,04,06,08,10,12,14,16,18,20
```

The total size of this array is 20B (each number is encoded into a 16b value). The individual values can be specified using any valid constant syntax, which can be mixed. To declare an array of characters (a string), one can use

```
MESSAGE    DC.B 'H','e','l','l','o',' ','W','o','r','l','d','!'
```

This array is 12 bytes long since there are 12 characters. The same result could be obtained with

```
MESSAGE    DC.B "Hello World!"
```

Note that the latter definition does not NULL-terminate the string, like some high level languages do. To declare a NULL-terminated string, one must explicitly include the zero

```
MESSAGE    DC.B "Hello World!",0  ;NULL Terminated
```

To create an uninitialized table or array, DS is used to reserve the total bytes for the array; to get the total byte count to be reserved one must multiply the number of bytes per element by the total number of elements needed. For example, to declare an uninitialized array of ten 8b values, one could use

```
BYTEARRAY    DS 10
```

On the other hand, an array of ten 16b values would be declared using

```
WORDARRAY    DS  20
```

To improve the ability to maintain the program, array lengths can be defined with EQU pseudo-ops, which are used to declare the array. For example, if the length is defined as

```
ARRAYLENGTH    EQU 10
```

Then the previous two array declarations could be written as

```
BYTEARRAY    DS ARRAYLENGTH
WORDARRAY    DS 2*ARRAYLENGTH
```

To access an array using extended or direct addressing, one can use assembler expressions to compute the address of the desired element. The index of the desired element $(0,1,2,\ldots)$ is multiplied by the element size and added to the address of the array to compute the address of the element. For example, to load BYTEARRAY[2] into A and WORDARRAY[3] into HX, one could use

```
LDA BYTEARRAY+2         ;load BYTEARRAY[2]
LDHX WORDARRAY+(3*2)    ;load WORDARRAY[3]
```

Each of the above instructions uses direct (or extended) addressing, depending on the locations of the arrays.

Offset-indexed addressing can be used in a similar way, but the addition is done as part of the effective address computation performed by the CPU. If HX points to the start of the array, then the offset is the index of the desired element multiplied by the element size. For example,

```
LDHX #BYTEARRAY    ;HX points to BYTEARRAY
LDA 2,X            ;load BYTEARRAY[2]
```

Note that the immediate notation in the instruction LDHX #BYTEARRAY is necessary to ensure that the value of the label BYTEARRAY is loaded into HX as an immediate; without the #, the extended addressing would be meant by the syntax.

Because indexed addressing can use a 16b offset, one can swap the roles of the offset and index, placing the index into HX and using the starting address of the ARRAY as the offset. For example,

```
LDHX #2            ;HX contains offset
LDA BYTEARRAY,X    ;load BYTEARRAY[2]
```

The advantage of this form is that the program can step through an array by changing the offset contained in HX. HX can also hold a variable index.

To create other data structures requires little additional effort. Here, the use of EQU pseudo-ops can simplify the programming and improve readability of the code. Suppose that a data structure called TIME, for example, is needed to hold a date and time, containing month (1B), day (1B), year (2B), hour (1B), minute (1B), and seconds (1B). EQU pseudo-ops can be used to define the size and offsets of the individual elements in the structure, as in

```
TIME_SIZE    EQU 7    ;number of bytes in TIME structure
TIME_MONTH   EQU 0    ;offset of month
TIME_DAY     EQU 1    ;offset of day
TIME_YEAR    EQU 2    ;offset of year
TIME_HOUR    EQU 4    ;offset of hour
TIME_MINS    EQU 5    ;offset of minutes
TIME_SECS    EQU 6    ;offset of seconds
```

To define initialized and uninitialized time structures, called TIME_START and TIME_END, one could then write

```
TIME_START   DC.B 01,01         ;01/01/2006 at 00:00:00
             DC.W 2006
             DC.B 0,0,0
TIME_END     DS   TIME_SIZE     ;an empty time structure
```

Notice that separate DC.B and DC.W pseudo-ops have been used, with a label only assigned to the first. The assembler will place the elements consecutively in the object code, organizing them collectively into a single memory structure. To load the seconds from TIME_START into accumulator A, for example, one could write

```
LDA    TIME_START+TIME_SECS    ; load seconds
```

When assembled, the expression TIME_START+TIME_SECS is converted to an address, which has offset +6 from TIME_START. The load would use extended or direct addressing.

3.4.2 IMPLEMENTING MATHEMATICAL FUNCTIONS VIA TABLES

Mathematical functions, such as sine and square root, are sometimes needed in typical microcontroller applications. Although efficient algorithms exist to compute these functions, the overhead of these algorithms is often too high. In many cases, it is possible to pre-compute the values of the function and store them in a constant table. During program execution, the function is then implemented by performing a look-up into the constant table implementing the function; the index into the table is based on the value of the function input.

The size of the table is determined by the precision of the input and outputs of the function. For example, if a function of an 8 bit input value is required then the table must contain $2^8 = 256$ entries, holding one pre-computed function value for each input. The size of each table entry is determined by the precision of the output needed. An example of a table of pre-computed values for the square root function (8b) is shown below. This table would be placed in the program constant variable section.

```
SQRT   DC.B    00,01,01,01,02,02,02,02,02,03,03,03,03,03,03,03
       DC.B    04,04,04,04,04,04,04,04,04,05,05,05,05,05,05,05
       DC.B    05,05,05,05,06,06,06,06,06,06,06,06,06,06,06,06
       DC.B    06,07,07,07,07,07,07,07,07,07,07,07,07,07,07,07
       DC.B    08,08,08,08,08,08,08,08,08,08,08,08,08,08,08,08
       DC.B    08,09,09,09,09,09,09,09,09,09,09,09,09,09,09,09
       DC.B    09,09,09,09,10,10,10,10,10,10,10,10,10,10,10,10
       DC.B    10,10,10,10,10,10,10,10,10,11,11,11,11,11,11,11
       DC.B    11,11,11,11,11,11,11,11,11,11,11,11,11,11,11,11
       DC.B    12,12,12,12,12,12,12,12,12,12,12,12,12,12,12,12
       DC.B    12,12,12,12,12,12,12,12,12,13,13,13,13,13,13,13
       DC.B    13,13,13,13,13,13,13,13,13,13,13,13,13,13,13,13
       DC.B    13,13,13,13,14,14,14,14,14,14,14,14,14,14,14,14
       DC.B    14,14,14,14,14,14,14,14,14,14,14,14,14,14,14,14
       DC.B    14,15,15,15,15,15,15,15,15,15,15,15,15,15,15,15
       DC.B    15,15,15,15,15,15,15,15,15,15,15,15,15,15,15,15
```

To access this table, one could use offset indexed addressing, with the (zero-extended) function input in HX and using the address of the table as the offset. The code sequence below shows the instructions necessary to compute $R = \sqrt{P}$.

```
LDX    P           ; X contains the value of P

CLRH               ; zero extend into H
LDA    SQRT,X      ; look-up table value
STA    R           ; store result
```

3.5   PROGRAMMING COMMON CONTROL STRUCTURES

High-level languages provide a diverse set of control structures such as loops and selection structures. In assembly language, these high-level structures are implemented with branch or loop primitive instructions. Structured languages use different means to prevent *spaghetti code*, unstructured code that is difficult to follow due to haphazard use of branches. In assembly language, there is nothing to prevent spaghetti code except disciplined programming. The programmer should strive to create programs that are well-organized based on the basic structures describe in this section.

3.5.1   SELECTION/DECISION STRUCTURES

A selection structure allows the program flow to be changed based on the value of a variable or an expression. The simplest selection structure is an IF-THEN-ELSE selection, which can be written in flowchart and pseudo-code notation as shown in Figure 3.4. This is a binary decision structure that can choose between two alternative code sections; the flows of the two sections eventually merge (even though this merge can be at the end of the program).

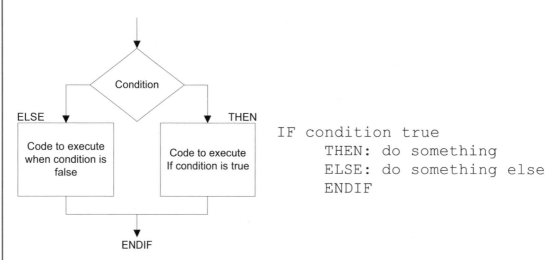

Figure 3.4: Flowchart and pseudo-code of a basic selection structure.

The condition in Figure 3.4 is a Boolean condition that results in true or false. Depending on the complexity of the expression given as the condition and on the available branch mnemonics, the condition might need to be evaluated before the branch. In the flowchart representation, the labels outside of the boxes represent potential labels in the assembly language program. This is shown in the example flowchart in Figure 3.5, which contains a possible HCS08 assembly language implementation. Note that the only labels in the example code that are actually needed are ELSE and ENDIF (the labels that are actually used as branch targets); however, the other labels can improve readability. Recall that labels must be unique and cannot be reused; thus, the programmer is required to select

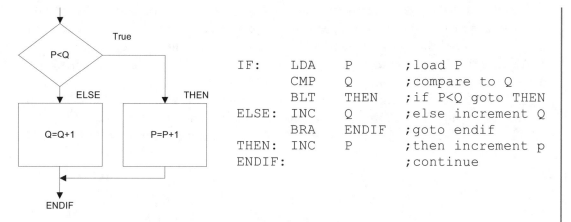

```
IF:     LDA     P          ;load P
        CMP     Q          ;compare to Q
        BLT     THEN       ;if P<Q goto THEN
ELSE:   INC     Q          ;else increment Q
        BRA     ENDIF      ;goto endif
THEN:   INC     P          ;then increment p
ENDIF:                     ;continue
```

Figure 3.5: Example IF-decision flowchart and a potential HCS08 assembly implementation.

different labels for each IF statement. The BRA ENDIF is necessary to skip over the THEN block of instructions after the ELSE is executed. In this form, which we will refer to as the BRANCH-TO-THEN form, the THEN section is at the branch target and the ELSE section at the branch fall-through.

The programmer can reorder the ELSE and THEN sections in memory by using the complement of the branch. In this BRANCH-TO-ELSE form, the ELSE block is at the branch target and the THEN block at the fall-through. This is shown in Figure 3.6. Notice that in the branch-to-ELSE form, the complement of BLT, BGE, is used.

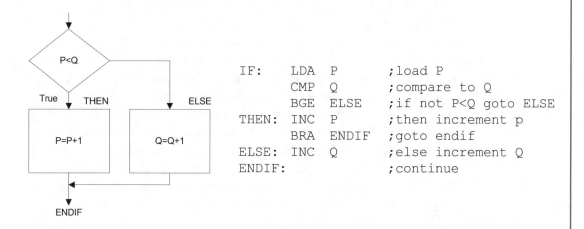

```
IF:     LDA   P          ;load P
        CMP   Q          ;compare to Q
        BGE   ELSE       ;if not P<Q goto ELSE
THEN:   INC   P          ;then increment p
        BRA   ENDIF      ;goto endif
ELSE:   INC   Q          ;else increment Q
ENDIF:                   ;continue
```

Figure 3.6: Alternate assembly implementation of IF-decision structure in Figure 3.5.

Both branch-to-THEN and branch-to-ELSE assembly language forms have similar execution times and code sizes and are thus equally acceptable. Notice that the second form results in a more natural program ordering from a code readability standpoint (the THEN block code precedes the ELSE block code); however, using the complement of the branch condition can sometimes lead to confusion. This is where well-commented code can help.

If the ELSE section is empty, the BRANCH-TO-ELSE form results in smaller code size. This is illustrated in Figure 3.7, which shows two methods for implementing an IF statement that has no ELSE. The BRANCH-TO-ELSE becomes BRANCH-TO-ENDIF and results in smaller and slightly faster code.

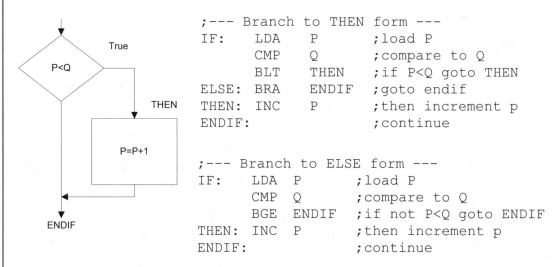

```
;--- Branch to THEN form ---
IF:     LDA     P       ;load P
        CMP     Q       ;compare to Q
        BLT     THEN    ;if P<Q goto THEN
ELSE:   BRA     ENDIF   ;goto endif
THEN:   INC     P       ;then increment p
ENDIF:                  ;continue


;--- Branch to ELSE form ---
IF:     LDA  P          ;load P
        CMP  Q          ;compare to Q
        BGE  ENDIF      ;if not P<Q goto ENDIF
THEN:   INC  P          ;then increment p
ENDIF:                  ;continue
```

Figure 3.7: Flowcharts and assembly code for two implementations of IF without an ELSE.

A switch statement also allows program flow to be controlled by an expression, where the expression is an integral variable type allowing for more than two outcomes. The value of the expression is compared against a set of constants, each of which can be assigned to a particular code section. A flowchart and pseudo-code are shown in Figure 3.8. Once the value of the expression is computed, it is compared to a list of constants in turn. When a match is found, the corresponding code section is executed. Each CASE section denotes a particular constant value Ci for the expression followed by its corresponding program section. Depending on the size of the variable used to hold the outcome of the expression and the number of cases of interest, the precision can exceed the number of cases; a *default case* is often included to handle this possibility. Multiple outcomes can also map to a single program section. The BREAK at each section indicates that the section has ended; otherwise, the meaning is that execution continues from one section to the next.

A switch structure can be organized in several ways. One form is shown in Code Listing 3.12, which shows an example code sequence that performs an arithmetic operation on a variable R based

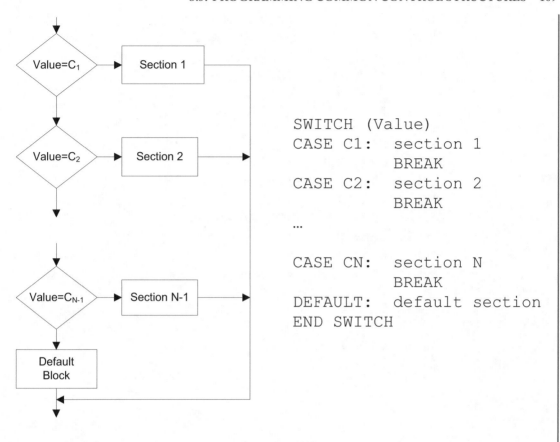

```
SWITCH (Value)
CASE C1:   section 1
           BREAK
CASE C2:   section 2
           BREAK
...

CASE CN:   section N
           BREAK
DEFAULT:   default section
END SWITCH
```

Figure 3.8: Flowchart of a typical switch selection structure.

on the value of a character constant. In this form, the cases are checked in turn with a series of CMP instructions followed by BEQ instructions; the target of each branch is the section that is executed if there is a match and the fall-through is the next case to be checked. The code sections then follow, with BREAK statements implemented by branching to the end (BRA ENDSW). Note that this structure will only work if all branch targets are within 128 bytes of their branches. In this form, the CMP and BEQ sequences can be replaced with a single CBEQA instruction.

An alternative implementation for the selection structure uses the complement branch, BNE, to jump-over code sections for cases that do not match until a match is found. This "leap-frog" approach is shown in Code Listing 3.13. While this form better mimics the natural flow of the switch statement, it is somewhat awkward when it is necessary to implement multiple cases mapping to the same section (see CASE_MP in the example). When multiple cases map to the same code section, all

```
 1   SWITCH:   LDA    C              ;compute expr
 2             CMP    #'+'           ;check CASE '+'
 3             BEQ    CASE_PL
 4             CMP    #'-'           ;check CASE '-'
 5             BEQ    CASE_MI
 6             CMP    #'/'           ;check CASE '/'
 7             BEQ    CASE_DV
 8             CMP    #'*'           ;check CASE '*'
 9             BEQ    CASE_MP
10             CMP    #'x'           ;check CASE 'x'
11             BEQ    CASE_MP
12             BRA    DEFAULT        ;DEFAULT CASE
13   ;---------- Code Sections -----------
14   CASE_PL:  INC    R
15             BRA    ENDSW          ;BREAK
16   CASE_MI:         DEC       R
17             BRA    ENDSW
18   CASE_DV:         ASR       R
19             BRA    ENDSW
20   CASE_MP   ASL    R
21             BRA    ENDSW
22   DEFAULT:  CLR    R;
23   ENDSW
```

Code Listing 3.12: HCS08 Implementation of Switch Structure.

of the cases must be checked before jumping over the code section. The code size is approximately the same for either implementation.

When the cases are positive integers from 0 to N, a jump table can be used to eliminate the successive CMP instructions. A jump table is an array of jumps to code sections. A jump table implementation of a switch statement is shown in Code Listing 3.14. In the example, there are 5 values for the switch expression, 0 through 4. The expression is computed (loaded) and compared to be within range (lines 1 and 2); if in range (line 3), the jump table is used. If out of range, the structure jumps to the default section (line 4).

The jump table (beginning on line 10) is an "array" of JMP instructions, each jumping to one code section of the switch statement. To use the jump table, the CPU must first jump to one of the instructions in the table, typically using offset indexed addressing to access this array. If I is the index of the jump to be taken from the table, them the effective address needs to be computed as JUMPTABLE+3*I (3 is the size of a JMP instruction using extended addressing). Using the addressing

```
1  SWITCH:    LDA  C            ;compute expr
2  CASE_PL:   CMP  #'+'         ;check CASE '+'
3             BNE  CASE_MI      ;
4             INC  R
5             BRA  ENDSW        ;BREAK
6  CASE_MI:   CMP  #'-'         ;check CASE '-'
7             BNE  CASE_DV
8             DEC  R
9             BRA  ENDSW
10 CASE_DV:   CMP  #'/'         ;check CASE   '/'
11            BNE  CASE_MP
12            ASR  R
13            BRA  ENDSW
14 CASE_MP    CMP  #'*'         ;check CASE '*'
15            BEQ  DO_MP
16            CMP  #'x'         ;check CASE 'x'
17            BNE  DEFAULT
18 DO_MP      ASL  R
19            BRA  ENDSW
20 DEFAULT:   CLR  R;
21 ENDSW
```

Code Listing 3.13: Alternate HCS08 Implementation of Switch Structure.

mode form where the base address of the jump table is used as the offset (JUMPTABLE, X), HX needs to contain 3*I. The index I is multiplied by 3 (lines 6 and 7), zero extended and transferred to HX (lines 8 and 9). The actual jump into the jump table is on line 10. The remainder of the code contains the case sections of the switch statement.

3.5.2   LOOP STRUCTURES

Loop structures implement algorithm repetition. There are three basic parts of a program loop: loop initialization, loop continuation/termination condition, and loop body. Loop initialization includes any operations that are needed before program control enters the loop, generally allowing variables to be initialized as necessary. The loop continuation/termination condition determines when the loop terminates; the test can be written using either sense: the loop continues if true (continuation) or terminates when false (termination). The loop body consists of the code that needs to be repeated. There are two basic loop forms, depending on whether the continuation/termination condition is done at the start or end of the loop. These two forms are shown in Figure 3.9.

```
 1  SWITCH:        LDA   I            ;get expression
 2                 CMP   #5           ;5 cases, 0 through 4
 3                 BLO   DO_JUMP      ;index in range-use jump table
 4                 JMP   DEFAULT      ;else execute default case
 5
 6  DO_JUMP:       LSLA               ;multiply by
 7                 ADD   I            ; size of jump table entry (3)
 8                 TAX
 9                 CLRH               ;zero-extend into HX
10                 JMP   JMPTABLE,X   ;jump to the case
11
12  JMPTABLE:      JMP   CASE_0       ;the jump table
13                 JMP   CASE_1
14                 JMP   CASE_2
15                 JMP   CASE_3
16                 JMP   CASE_4
17
18  CASE_0:        ;insert case 0 code
19                 JMP   ENDSW
20  CASE_1:        ;insert case 1 code
21                 JMP   ENDSW
22  CASE_2:        ;insert case 2 code
23                 JMP   ENDSW
24  CASE_3:        ;insert case 3 code
25                 JMP   ENDSW
26  CASE_4:        ;insert case 4 code
27                 JMP   ENDSW
28  DEFAULT:       ;insert default case code
29                 JMP   ENDSW
30  ENDSW:
```

Code Listing 3.14: HCS08 Program Illustrating Use of a Jump Table.

Loops whose continuation/termination condition evaluation is done at the top of the loop are sometimes referred to as WHILE loops while those with the evaluation performed at the end of the loop are referred to as DO-WHILE loops. The basic difference between the two loop forms is that the loop body in the DO-WHILE form is always executed for at least 1 iteration; while the WHILE loop may iterate 0 times. The DO-WHILE form is useful when the initialization code and loop body are the same, allowing them to be easily combined.

The WHILE form requires two branches to implement: a conditional branch to test the termination/continuation condition at the top of the loop, and a BRA to loop back at the bottom of the

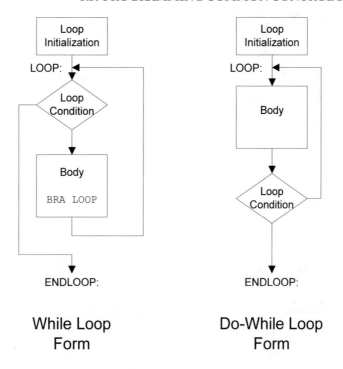

Figure 3.9: While and do-while loop flowcharts.

loop. In the DO-WHILE form, the conditional branch at the bottom of the loop serves as both the looping branch and the termination/continuation condition branch.

Consider the example in Figure 3.10, which shows a sequence of instructions that finds the largest multiple of the unsigned 8b integer P that is less than 256, assuming P is non-zero. The result is stored in R. One possible solution is to repeatedly accumulate P until the result overflows; thus, the loop termination is based on unsigned overflow. Two possible loop structures are shown in Figure 3.10. The While form on the left side of Figure 3.10 tests the termination condition (C=1) at the top of the loop; thus it requires the extra step of clearing the C flag during the loop initialization. The Do-While form on the right side of Figure 3.10, however, tests the loop condition at the bottom of the loop. Since the carry flag is based on the instructions in the loop body, there is no need to clear the carry during loop initialization. Also, the loop continuation branch serves as the looping branch.

The assembly language code for both forms is shown in the Code Listing 3.15. Clearly, in this case the DO-WHILE form is superior: not only does it have smaller machine code, but there are only two instructions repeated in the loop: ADD and BCC. In the WHILE form, there are three instructions repeated in the loop (BCS,ADD,BRA) which results in more time to execute. To decide which form to

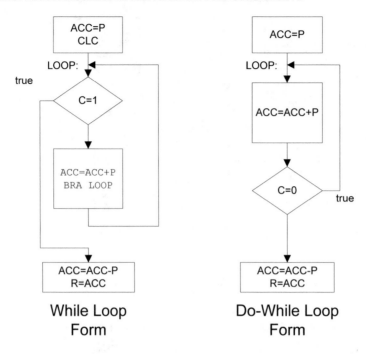

Figure 3.10: Flowchart of a loop implemented using While and do-while forms.

use, simply answer the question "is it possible for the loop to iterate zero times?" If the answer is yes, the WHILE form is required (to test the condition at the top of the loop). Otherwise, the DO-WHILE form is likely to be more efficient.

Counter-based loops can be implemented with WHILE or DO-WHILE forms, as shown in Figure 3.11. Generally, the counter is initialized in the loop initialization section and updated in the loop body. The loop termination condition is based on the terminal value of the counter. As it is for general loops, the WHILE form should be used if there is a possibility of zero iterations in a counter loop.

The loop primitive instructions (DBNZ, DBNZA, DBNZX) perform a counter update followed by a branch if not equal (to zero). They are useful for creating counter loops of the DO-WHILE form that count down from N to 1. If a counter loop is needed where the loop body is executed N times for (counter values from N to 1), then a loop primitive instruction is the most efficient choice.

```
A.
1    ;WHILE Form
2    INIT:          LDA     P       ;ACC=P
3                   CLC
4    LOOP:          BCS     END  ;test at top
5    BODY:          ADD     P
6                   BRA     LOOP
7    END:           SUB     P
8    STA            R
B.
1    ;DO-WHILE Form
2    INIT:          LDA     P       ;ACC=P
3
4
5    LOOP:          ADD     P
6                   BCC     LOOP ;test at end
7    END:           SUB     P
8                   STA     R
```

Code Listing 3.15: HCS08 Implementations of Flowcharts in Figure 3.10.

3.6   SUBROUTINES

Subroutines have several uses in programming. Subroutines allow programs to be modularized, which facilitates maintenance, collaboration and organization of code. Subroutines also allow code reuse, where repeated sequences of instructions can be replaced by a single sequence, with inputs and outputs parameterized, and used in several places in a program. This facilitates code maintenance, since there is only a single copy of each subroutine to keep up to date. It also results in smaller programs, although execution time slightly increases due to the overhead of passing parameters to/from the subroutines in addition to the subroutine call overhead itself. Due to the well-organized programs that result, subroutines are often used even when they are called only a single time during program execution.

A subroutine can be called from multiple points within a program. In order to return back to the point from which the subroutine was called, the return address needs to be saved. The return address always points to the instruction following the one that called the subroutine. Because subroutines can be called from subroutines (nesting), it is required to be able to save an unknown number of return addresses at any given time. Thus, the return address is pushed onto the stack at each subroutine call.

The HCS08 CPU has three instructions that are specific to subroutine implementations: BSR, JSR, and RTS. BSR and JSR behave like BRA and JMP instructions, except each one automatically pushes the return address (the address of the instruction immediately after the BSR or JSR) onto the

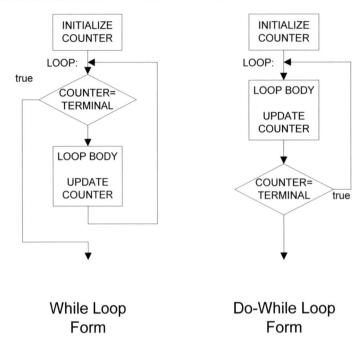

Figure 3.11: Flowcharts of alternate implementations of counter-based loops.

stack. This causes program control to branch or jump to the subroutine code. When the subroutine is complete, the RTS instruction pulls the return address off the stack into the program counter, effecting a jump back to the return address; in essence, the RTS is equivalent to "PULL into PC."

The subroutine call process is illustrated in Figure 3.12. Before the BSR is executed by the CPU, control is in the calling program (the caller). Immediately after the BSR, the return address is at the top of the stack and the PC is pointing to the first instruction in the subroutine. Until an RTS instruction is executed (or a nested subroutine call is issued) program control remains in the subroutine.

The figure shows a similar view of memory just before the RTS is executed and immediately after. The RTS pulls the return address off the stack and places it in the PC; the stack pointer is therefore restored to the point it was at before the subroutine was called (the calling programs stack structure is the same as it was before the call). It is extremely important for the programmer to ensure that the stack is cleaned up after the subroutine is complete, to guarantee that the return address is at the top of the stack. Because the return address is the address of the instruction immediately after the subroutine call, the PC is pointing to the instruction after the BSR. Thus, control is returned to the calling program.

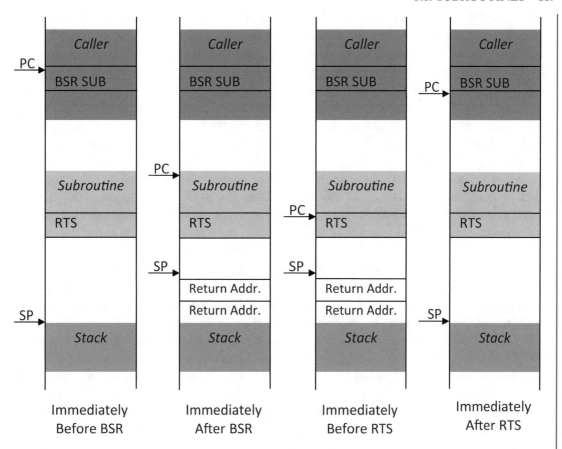

Figure 3.12: Memory map of stack at different stages of subroutine call sequence.

3.6.1   REGISTER SAVING

When program control is temporarily passed to the subroutine, the subroutine can modify memory and registers. There must be coordination between the calling program and the subroutine about what registers can be modified and how. Generally, the calling program will be using registers for holding operands, but the subroutine also needs to use these registers to hold its operands. Instead of coordinating the sharing of the registers, their values can be saved on the stack before the subroutine modifies them and restored before control is returned to the calling program. This register saving is not automatic and must be programmed with push and pull instructions.

Register saving can be done by the calling program (referred to as caller-saving) or by the subroutine (referred to as callee-saving). With caller-saving, the assembly language instructions to do the register saving must be included at each point in the program from which the subroutine is called, generally resulting in larger programs. Callee-saving has the advantage of having the register saving code written once in the subroutine itself, resulting in more compact programs. In addition, the subroutine needs only to save those registers that it will modify, which eliminates saving registers that will not be modified (with caller saving, the caller would save all registers in use whether modified by the subroutine or not). Having the register saving at a single point within the subroutine means that if the code is modified, every call to the subroutine does not need to be changed, facilitating code maintenance. In some instances, callee-saved registers might contain values not being used by the caller and thus the register saving is done unnecessarily. The simplicity and code maintenance advantages of callee-saving, however, generally outweigh this slight overhead. While it is possible to use both callee-saving and caller-saving for different subroutines in the same program, generally one type of register saving is used for consistency. Compilers of high-level languages can perform such optimizations automatically; for the assembly language programmer consistency, simplicity and shortened development and debug time often outweigh small savings in code size and execution time.

If caller-saving is used, it is *bad* programming practice to omit register saving code for caller-saved registers that are known to be unused by a subroutine; this is because if the subroutine is later modified, the programmer will need to go back and insert the caller-saving code for any newly modified registers at each point in the program where the subroutine is called. This can easily lead to code maintenance headaches and bugs. When callee-saving is used, it is relatively simple to modify the callee-saving code is if the subroutine is modified. It is therefore good programming practice to caller-save all registers currently in use by the calling program, independent of how the subroutine will use the registers. It is also acceptable to callee-save only the registers that a subroutine modifies.

A skeleton program demonstrating caller-saving is provided in Code Listing 3.16. The stack is initialized on lines 9 and 10, a necessary step whenever subroutines are used because subroutine calls always involve the stack. The subroutine call begins on line 14. The CPU registers are pushed onto the stack before the subroutine call; any registers not being used by the calling program at the time of the call can be omitted from the register saving. Notice that when caller-saving is used, all of the work required to implement register saving is performed at the point of the subroutine call; register saving does not have to be done in the subroutine. Also notice that the registers are pulled from the stack in reverse order that they were pushed after the subroutine call.

A skeleton program illustrating callee-saving of registers is shown in Code Listing 3.17. The first instructions in the subroutine callee-saves the registers that will be modified by the subroutine. Immediately before the RTS, the original register values are restored, in reverse order. It is okay to omit the callee-saving code for registers that will not be modified by the subroutine. It is not possible to determine whether or not the calling program needs the values that are being saved, because a subroutine can be called form many points within a program.

```
1   ;----------------------------------------------------------
2   ;Skeleton Program to Illustrate Use of Caller Saved Registers
3   ;----------------------------------------------------------
4   RAMSTART        EQU     $0060       ;Address of Start of RAM
5   RAMSIZE         EQU     256         ;Size of RAM, in Bytes
6   ROMSTART        EQU     $F000       ;Start address of ROM
7
8                   ORG     ROMSTART
9                   LDHX    #{RAMSTART+RAMSIZE}
10                  TXS                     ;initialize SP
11  START:          ;program code starts here
12
13  ;--------------subroutine call begins here-----
14                  PSHA            ;caller save registers
15                  PSHX
16                  PSHH
17                  BSR     SUBRTN  ;call subroutine
18                  PULH            ;restore registers
19                  PULX
20                  PULA
21  ;--------------subroutine call ends here-----
22
23                  ;program code continues here
24                  BRA     START
25
26  ;------------------------------------------------
27  ;Subroutine skeleton SUBRTN
28  ;------------------------------------------------
29  SUBRTN:         ;insert subroutine code here
30                  RTS     ;return to caller
```

Code Listing 3.16: Program Illustrating Caller Saved Registers.

3.6.2 STACK VARIABLES

The stack is often used for storing the intermediate results of calculations and as a place to temporarily hold data when there are not enough CPU registers. The stack can also be used as a place to create variables used by subroutines. As noted above, the variables required by a subroutine are often only needed while the subroutine is active – their values are not retained across subroutine calls. In addition, nesting and recursion can create a need for multiple simultaneous copies of a subroutine's

```
1  ;-----------------------------------------------------------
2  ;Skeleton Program to Illustrate Use of Callee Saved Registers
3  ;-----------------------------------------------------------
4  RAMSTART EQU  $0060   ;Address of Start of RAM
5  RAMSIZE  EQU  256     ;Size of RAM, in Bytes
6  ROMSTART EQU  $F000   ;Start address of ROM
7
8           ORG  ROMSTART
9           LDHX       #{RAMSTART+RAMSIZE}
10          TXS        ;initialize SP
11 START:   ;program code starts here
12
13 ;--------------subroutine call begins here-----
14          BSR SUBRTN ;call subroutine
15 ;--------------subroutine call ends here-----
16
17          ;program code continues here
18          BRA  START
19
20 ;-------------------------------------------------
21 ;Subroutine skeleton SUBRTN
22 ;-------------------------------------------------
23 SUBRTN:  PSHA    ;callee save registers
24          PSHX
25          PSHH
26          ;insert subroutine code here
27          PULH    ;restore registers
28          PULX
29          PULA
30          RTS   ;return to caller
```

Code Listing 3.17: Program Illustrating Callee-Saved Registers.

variables. For these reasons, it is generally more effective to temporarily allocate the memory needed for local variables from the system stack, freeing up that memory when the subroutine returns.

Creating stack variables is simply a matter of creating the necessary amount of memory space on the top of the stack, then logically mapping variables into this space. The principle difference between using stack variables and globally named variables is that the program cannot use direct or extended addressing to address them; that is, it cannot reference the variables by name (address) because data access on the stack is always relative to the current location of the top of the stack. Therefore, stack variables must be accessed through indexed or offset indexed addressing, using either the stack pointer or index register as the base.

The Code Listing 3.18 provides an example of allocating and initializing stack variables. The

```
1   ;--------------------------------------------------
2   ;Subroutine example demonstrating use of stack variables
3   ;Three 8b stack variables define: VI, VJ, and VK
4   ;--------------------------------------------------
5   SUBRTN:         PSHA                    ;callee save registers
6                   PSHX
7                   PSHH
8
9                   ;Create space for stack variables
10                  AIS    #-3              ;allocate 3 bytes on stack
11                  ;initialize stack variables VI, VJ
12                  CLR    1,SP             ;VI=0
13                  LDA    #32
14                  STA    2,SP             ;VJ=32
15                  CLR    3,SP             ;VK=0
16
17  ;subroutine body inserted here
18                  AIS #3                  ;deallocate stack variables
19                  PULH                    ;restore registers
20                  PULX
21                  PULA
22                  RTS    ;return to caller
```

Code Listing 3.18: Program to Illustrate Use of Stack Variables.

subroutine creates three 8b stack variables, called VI, VJ, and VK. After callee-saving the registers, 3 bytes are allocated on the top of the stack on line 10. For the remainder of the subroutine, these three bytes can be accessed relative to the stack pointer. The programmer is free to allocate the bytes to the variables; in the example, the byte on top of the stack (1,SP) is assigned as VI, followed by VJ (2,SP) and VK (3,SP). Example code showing how these variables can be initialized is shown on lines 12 through 15. Each variable is accessed using stack pointer offset addressing. When the subroutine is finished with the variables, but before it restores the callee saved registers, the space allocated to the stack variables must be released. This is accomplished on line 19 with the AIS instruction; the values can also be pulled if they are to be used as return values for the subroutine.

There are two possible difficulties that can arise when using stack variables. One is that the SP can move during the execution of the subroutine if the subroutine uses the stack to hold temporary data, requiring the programmer to keep track of how far SP has moved relative to the stack variables and adjust the offsets accordingly. To address this problem, the programmer can define a pointer to them by copying the stack pointer to HX after creating the stack variables; offset indexed addressing can then used to access the data. The second difficulty is the added complexity of having to refer to variables by offset rather than by name (label). One solution is to use EQU pseudo-ops to define the variable offsets, then use the labels in offset indexed addressing.

Both of these modifications have been made to the subroutine from Code Listing 3.18, and the modified subroutine code is shown in Code Listing 3.19. On lines 9 through 11, EQU pseudo-ops are used to define offsets for VI, VJ, and VK. After allocating space for stack variables on line 13, SP is copied to HX with TSX, which makes HX point one byte up from SP in memory so that HX points to VI; this is why the offsets declared for VI, VJ, and VK are 0, 1, and 2 (instead of offsets 1, 2, and 3 used in stack-offset indexed addressing). The defined offset labels can be used to access the variables using the notation in lines 16 and 18; with the offset labels, the programmer no longer needs to keep track of how variables have been structured on the stack, which makes the code easier to read and maintain. Should the main body of the subroutine use the stack for other purposes, HX will always provide a fixed reference point to access stack variables. If HX is needed for other purposes in the subroutine, it can be stored in a local variable or itself temporarily pushed onto the stack.

3.6.3   PASSING PARAMETERS AND RETURNING RESULTS

The calling program and subroutine do not always need to exchange data between them. In many cases, however, subroutines require input parameters from the calling program and return results back to the calling program. There are four basic methods by which data can be exchanged: registers, flags, global variables and the stack.

The simplest method for passing data between the subroutine and calling program is via the CPU registers. In the HCS08 CPU, this includes H, X, and A (SP and PC have dedicated use). Any of these can be used to pass 8b operands, while HX can also be used to pass 16b operands or pointers to the subroutine. When using a register to pass data from a subroutine to calling program, that register must be omitted from any register saving; otherwise, the restore of the register will overwrite the value of the result to be returned.

Using CPU flags to pass parameters allows exchange of Boolean data. Because the carry flag can be explicitly set and cleared with opcodes SEC and CLC, it is often chosen as the register to use for this purpose. Any of the CPU flags could be used, although manipulating them requires additional programmer effort.

```
1   ;---------------------------------------------------
2   ;Subroutine example demonstrating use of stack variables
3   ;Three 8b stack variables define: VI, VJ, and VK
4   ;---------------------------------------------------
5   SUBRTN:   PSHA     ;callee save registers
6             PSHX
7             PSHH
8             ;define offsets of stack allocated variables
9   VI        EQU  0
10  VJ        EQU  1
11  VK        EQU  2
12            ;Create space for stack variables
13            AIS  #-3    ;3 bytes
14            TSX    ;HX points to stack variable on top
15            ;initialize stack variables VI, VJ
16            CLR  VI,X    ;VI=0
17            LDA  #32
18            STA  VJ,X    ;VJ=32
19            CLR  VK,X
20  ;subroutine code inserted here
21            AIS #3       ;deallocate stack variables
22            PULH    ;restore registers
23            PULX
24            PULA
25            RTS  ;return to caller
```

Code Listing 3.19: Subroutine Demonstrating Use of Stack Variables.

An example of a subroutine that uses registers and flags to exchange data is shown in Code Listing 3.20. In this example, the subroutine ISMULT8 determines if its 8b unsigned parameter, passed via register X, is evenly divisible by 8. The result is returned back to the calling program in the carry flag such that C=1 indicates that the number is divisible by 8. Callee-saving is used. For simplicity, only the actual subroutine call is shown (startup code and initialization code is omitted).

The subroutine call begins on line 6, which loads the parameter N into register X in order to pass it to the subroutine. The subroutine is then called with the BSR; upon returning from the call, the registers are all exactly as they were before the call. The carry flag, however, contains the result

```
1    ;-----------------------------------------------------------
2    ;Code to illustrate parameter passing in registers and flags
3    ;-----------------------------------------------------------
4
5    ;-------------subroutine call begins here-----
6                   ;call ISMULT8(N)
7                   LDX    N              ;load subroutine parameter
8                   BSR    ISMULT8        ;call subroutine
9                   ADCA   #0             ;accumulate result into Acc A
10   ;-------------subroutine call ends here-----
11
12
13   ;-------------------------------------------------
14   ;Subroutine ISMULT8(N), Returns C=1 if 8b unsigned
15   ;value N, passed in register X, is evenly divisibly by 8
16   ;-------------------------------------------------
17   ISMULT8:       PSHX                  ;callee save register(s)
18                  LSRX
19                  BCS    ISNOTDIV       ;is divisible by 2?
20                  LSRX
21                  BCS    ISNOTDIV       ;and again?
22                  LSRX
23                  BCS    ISNOTDIV       ;and again?
24   ISDIV:         SEC                   ;is divisible by 8, set C
25                  BRA    RTNISM8
26   ISNOTDIV       CLC                   ;not divisible, clear C
27   RTNISM8        PULX                  ;restore register(s)
28                  RTS                   ;return to caller
```

Code Listing 3.20: Parameter Passing Using Registers.

of the subroutine. This result can be accumulated in A (as shown here on line 8), rotated or shifted into A, or tested via a BCS or BCC instruction.

The subroutine code begins on line 16. The only register modified by the subroutine is X; this is callee-saved on line 16. No other registers are modified and thus they are omitted from register saving. Even though X is used to pass parameters to the subroutine, it does not pass anything back to the calling program; thus, it is callee-saved. The algorithm of the subroutine is to shift the operand in X right 3 times, checking each time if the carry flag is set (if the bit shifted out is a 1). If a 1 is detected, the number is not evenly divisible by 8 and a branch to ISNOTDIV is made, which clears the carry flag before returning. If all three bits shifted out are clear, then the number is divisible by 8 and the last branch falls through to the instruction at label ISDIV on line 23 (the label is not needed, but included to help document the code). For either outcome (ISDIV or ISNOTDIV) program control continues to a common return point on line 26 (label RTNISM8) which handles the restoring of the registers and the return call. Having a common return point facilitates code maintenance and helps the programmer to ensure that the stack is clean before the RTS executes (any data left on the stack by the subroutine would cause RTS to behave unpredictably).

Passing parameters via memory can use global (named) variables or the stack. Global variables make it difficult in all but the simplest programs to coordinate data exchange among nested subroutines; in addition, unless a global variable has some use in addition to passing parameters between subroutines, the memory space that it ties up is wasted when the subroutine is not being executed. Thus, exchange of data via global variables should only be performed when the global variables already exist for another purpose in the program. The programmer should not be lazy and create global new global variables that have no purpose other than subroutine parameter passing.

When passing parameters via registers does not provide sufficient space, the stack is the appropriate mechanism through which to pass them via memory. The difficulty in passing parameters via the stack is that the return address is placed on top of the stack when the subroutine is called. The return address must also be on top of the stack just before the subroutine returns, preventing the subroutine from placing additional return parameters on the stack. If it is necessary to return data from a subroutine via the stack, the calling program must create space on the stack before the subroutine call to hold these results. The subroutine can access this space and place return parameters in it.

When calling a subroutine, there can be several types of data on the stack related to the subroutine: parameters, return address, saved registers and local variables (Figure 3.13). Grouped together, this data is referred to as a stack frame. A stack frame can be viewed as a memory map of the area on the stack belonging to the subroutine call. Depending on whether callee or caller saving is used, the exact ordering of this information can differ. For callee-saved registers, the stack frame would have the structure shown in Figure 3.13.

All of the data in the stack frame is accessible by the subroutine; the calling program, on the other hand, only has access to the parameters; the remainder of the stack frame (return address, callee-saved registers, and local variables) doesn't exist before the subroutine call or after the RTS. The subroutine must clean the local variables and callee-saved registers off the stack before RTS to ensure correct program behavior. The steps performed by the calling program and the subroutine to setup and use the stack are summarized in Table 3.5.

The subroutine being called is shown in Code Listing 3.21. After callee-saving registers and establishing HX as the frame pointer on lines 6 through 9, labels are defined for the offsets of the parameters on the stack; these offsets take into account the return address and callee saved registers on the stack. The parameters are accessed to compute the result parameter (lines 15-24). The callee-saved registers are restored and the subroutine returns.

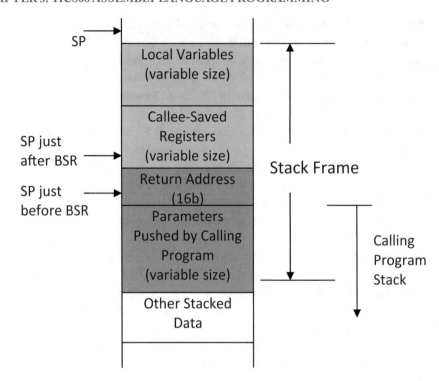

Figure 3.13: Memory-Map of Subroutine Stack Frame.

The example in Code Listing 3.22 illustrates a call to a subroutine that uses parameter passing via the stack. The subroutine, PROD32(VP,VI,VJ) (not shown), adds two 8b signed integers, VI and VJ, and multiplies the result by 32. The calling program is issuing the call PROD32(Z,U,V) to compute Z=32*(U+V). The parameters are pushed onto the stack on lines 3 through 6. Line 7 creates space on the stack into which the result can be placed by the subroutine. After the call to the subroutine, the result is pulled from the stack and stored to Z (lines 9 through 12). Finally, the parameters are removed from the stack on line 13.

3.7   ANALYSIS OF CODE SIZE AND EXECUTION TIME

When comparing two code sequences, one can consider the two goals of code size and execution time. Code size is static; it does not depend on the program state at run time. Therefore, code size can be determined by counting the machine code bytes necessary to encode an instruction sequence in machine code. Often, it is not necessary to determine the values of the machine code bytes; one can simply count the number of bytes that would be needed. For example, consider the two methods in Code Listing 3.23 for counting the number of bits that are set at memory location 0080_{16}. In the

Table 3.5: Sequence of steps needed to use stack for parameter passing.

Sequence	Calling Program Action	Subroutine Action
1	Place argument parameters on stack and create space for return parameters	
2	Issue BSR (which places return address on the stack)	
3		Callee-save registers
4		Set up stack variables
5		Execute main subroutine code, accessing parameters from the stack and storing results into result parameters on the stack
6		Clean local variables off stack
7		Restore registers
8		Issue RTS
9	Retrieve return parameters from stack	
10	Remove calling parameters to clean up stack	

comment next to each instruction is the number of bytes of machine code occupied by the instruction in memory. To determine the total program size one simply counts the total number of bytes (the actual values of the machine code bytes are not needed). The solution on the left, which uses a loop approach, results in a total of 1+2+1+2+1+2+2+1 = 12 bytes of machine code. The solution on the right, which does not use loops, results in a code size of 27 bytes of machine code.

Execution time can be measured in units of clock cycles or seconds. To compute the execution time in cycles, one must compute the number of times each instruction executes and multiply that by the number of clock cycles used by the CPU to execute the instruction. The same instruction sequences from Code Listing 3.23 are shown in Code Listing 3.24, this time with comments indicating the number of clock cycles required to execute each instruction and the number of times the instruction is executed (with the notation Nx indicating N times). In the loop form of the code on the left, the loop is executed 8 times, thus each of the instruction inside the loop is executed 8 times. Thus, the total execution time, in clock cycles, is $1 \times (1 + 3 + 2 + 2 + 3) + 8 \times (6 + 2 + 4) = 107$ *cycles*. For the code on the right, there are no loops and each instruction is executed only once. Thus, the total execution time is $1 \times (1 + 3 + 1 + 2 + 1 + 2 + 1 + 2 + 1 + 2 + 1 + 2 + 1 + 2 + 1 + 2 + 1 + 2) = 28$ *cycles*.

To convert from cycles to seconds, the execution time in seconds must be multiplied by the clock period, which has units of seconds per cycle. The clock frequency is usually known; thus, the execution time is computed as

```
 1   ;-------------------------------------------------
 2   ;Subroutine PROD32(VP,VI,VJ), returns in VP the 16b VP value
 3   ;VP=32*(VI+VJ) VI and VJ are 8b signed integers passed via stack.
 4   ;VP is passed via stack
 5   ;-------------------------------------------------
 6   ISMULT8:      PSHX                    ;callee-save register(s)
 7                 PSHA
 8                 PSHH
 9                 TSX                     ;establish frame pointer
10   ;There are 3 callee saved register bytes and 2 return address bytes
11   ;on top of the parameters.  Thus, offset to start of params is 5
12   VI            EQU     5+3             ;8b unsigned value
13   VJ            EQU     5+2             ;8b unsigned value
14   VP            EQU     5+0             ;16b return parameter
15                 LDA     VI,X            ;get parameter VI
16                 ADD     VJ,X            ;add VJ
17                 STA     VP+1,X          ;store LSB of sum
18                 LDA     #0              ;sign-extend into MSB
19                 SBC     #0              ;if C=1, MSB=FF, else MSB=0
20                 STA     VP,X            ;store MSB
21   MULTIPLY      LDA     #5              ;shift left 5 times
22   SHIFTIT       LSL     VP+1,X          ;shift left LSB
23                 ROL     VP,X            ;shift into MSB w/ carry-in
24                 DBNZA   SHIFTIT
25   RTNISM8       PULH                    ;restore register(s)
26                 PULA
27                 PULX
28                 RTS                     ;return to caller
```

Code Listing 3.21: Subroutine Using Parameter Passing via Stack.

$$\text{Execution Time } (s) = \text{number of cycles} \times \frac{1}{\text{clock frequency}}$$

For example, if the CPU clock were running at 3.2 MHz, then the execution times for the two sequences above would be $107 \times \frac{1}{3.2 \cdot 10^6} = 33.4375 \ \mu s$ and $28 \times \frac{1}{3.2 \cdot 10^6} = 8.75 \ \mu s$, respectively.

Notice that the loop form of the program is more efficient in terms of code size: 12 B for the loop form, 27 bytes for the non-loop form. On the other hand, the execution time for the loop form is almost 4 times as long, requiring 107 cycles instead of 28. It is generally the case that encoding a repetitive code sequence in loop form results in a smaller program but requires more time to execute due to the overhead of implementing the loop. Compilers often exploit this tradeoff when

```
1  ;--------------subroutine call begins here-----
2            ;call PROD32(Z,U,V)
3            LDA  U   ;get U
4            PSHA     ;push onto stack
5            LDA  V   ;get V
6            PSHA     ;push onto stack
7            AIS  #-2      ;create space for return param
8            BSR  PROD32  ;call subroutine
9            PULA     ;get result MSB
10           STA  Z   ;store it
11           PULA     ;get result LSB
12           STA  Z+1     ;store it
13           AIS  #2 ;remove parameters from stack
14 ;--------------subroutine call ends here-----
```

Code Listing 3.22: Call to Subroutine Using Stack Variables.

```
           CLRA          ; 1 bytes          CLRA          ; 1bytes
           LDX $80       ; 2 bytes          LDX   $80     ; 2bytes
           PSHX          ; 1 byte           LSLX          ; 1bytes
           LDX #$8       ; 2 bytes          ADC   #0      ; 2bytes
LOOP       LSL $01,SP    ; 1 byte           LSLX          ; 1bytes
           ADC #0        ; 2 bytes          ADC   #0      ; 2bytes
           DBNZX LOOP    ; 2 bytes          LSLX          ; 1bytes
           PULX          ; 1 byte           ADC   #0      ; 2bytes
                                            LSLX          ; 1bytes
                                            ADC   #0      ; 2bytes
                                            LSLX          ; 1bytes
                                            ADC   #0      ; 2bytes
                                            LSLX          ; 1bytes
                                            ADC   #0      ; 2bytes
                                            LSLX          ; 1bytes
                                            ADC   #0      ; 2bytes
                                            LSLX          ; 1bytes
                                            ADC   #0      ; 2bytes
```

Code Listing 3.23: Comparison of Code Size of Loop-Based and Unrolled Code.

```
          CLRA        ; 1 cycles, 1x          CLRA        ; 1cycles, 1x
          LDX $80     ; 3 cycles, 1x          LDX $80     ; 3cycles, 1x
          PSHX        ; 2 cycles, 1x          LSLX        ; 1cycles, 1x
          LDX #$8     ; 2 cycles, 1x          ADC #0      ; 2cycles, 1x
   LOOP   LSL $01,SP  ; 6 cycles, 8x          LSLX        ; 1cycles, 1x
          ADC #0      ; 2 cycles, 8x          ADC #0      ; 2cycles, 1x
          DBNZX LOOP  ; 4 cycles, 8x          LSLX        ; 1cycles, 1x
          PULX        ; 3 cycles, 1x          ADC #0      ; 2cycles, 1x
                                              LSLX        ; 1cycles, 1x
                                              ADC #0      ; 2cycles, 1x
                                              LSLX        ; 1cycles, 1x
                                              ADC #0      ; 2cycles, 1x
                                              LSLX        ; 1cycles, 1x
                                              ADC #0      ; 2cycles, 1x
                                              LSLX        ; 1cycles, 1x
                                              ADC #0      ; 2cycles, 1x
                                              LSLX        ; 1cycles, 1x
                                              ADC #0      ; 2cycles, 1x
```

Code Listing 3.24: Comparison of Execution Time of Loop-Based and Unrolled Code.

```
1                CLRA        ; 1cycles, 1x
2                LDX $80     ; 3cycles, 1x
3     LOOP:      LSLX        ; 1cycles, ?x
4                ADC #0      ; 2cycles, ?x
5                TSTX        ; 1cycles, ?x
6                BNE LOOP    ; 3cycles, ?x
```

Code Listing 3.25: Code Size and Execution Time of Nested Loop.

optimizing code for speed using a technique called loop unrolling. In loop unrolling, the number of loop iterations is reduced by repeating the loop body N times. For example, the loop form of the code above can be rewritten as shown in Code Listing 3.25 (with changes emphasized in bold font). The result of this code is that the loop body is executed only 4 times, but is double the size. Since the DBNZX instruction is executed 4 fewer times, the code requires 16 clock cycles fewer to execute $1 \times (1 + 3 + 2 + 2 + 3) + 4 \times (6 + 2 + 6 + 2 + 4) = 91$ cycles compared to 107 cycles for the original code. Of course, the code size has increased slightly due to this change. It is possible to exploit this execution time/code size tradeoff in other ways, but this is beyond the scope of this book.

Finally, as an example of code for which it is not possible to exactly determine the execution time, consider the loop provided in Code Listing 3.26. In this code, the number of iterations depends on the value in memory location 0080_{16}, which is only known at run time. The execution time can

```
1              CLRA            ; 1 cycles, 1x
2              LDX $80         ; 3 cycles, 1x
3              PSHX            ; 2 cycles, 1x
4              LDX #$4         ; 2 cycles, 1x
5    LOOP      LSL $01,SP      ; 6 cycles, 4x
6              ADC #0          ; 2 cycles, 4x
7              LSL $01,SP      ; 6 cycles, 4x
8              ADC #0          ; 2 cycles, 4x
9              DBNZX LOOP      ; 4 cycles, 4x
10             PULX            ; 3 cycles, 1x
```

Code Listing 3.26: Example of Code for which Execution Time cannot be Statistically Determined.

be expressed as $1 + 3 + iterations \times (1 + 2 + 1 + 3)$, but iterations can only be exactly determined if the value in location 0080_{16} is known. Although it may be possible to estimate the average value of *iterations*, this is beyond the scope of this text.

3.8   CHAPTER PROBLEMS

1. Convert each of the following to machine code

    a. NOP

    b. ASRA

    c. LDX #45

    d. LDX $80

    e. LDX $F000

    f. ADD ,X

    g. ADD 3,X

    h. MOV$80,$00

    i. BSET 3,$80

2. Assemble the following code by hand. Show the work.

```
        ORG $EE00
START   CLRA
        LDX #$08
LOOP    ASRA
        BCC SKIP
        INCA
SKIP    DBNZX  LOOP
END     BRA END
```

3. Write a skeleton program that is organized according to the following memory map. Put in comments indicating where program code and program data should be inserted. Also include the SP initialization code.

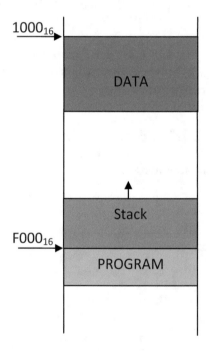

1000_{16}

DATA

Stack

$F000_{16}$

PROGRAM

4. Write the assembly pseudo-ops that

   a. Define a table of 8b constants named TABLE7 that contains the 16 elements 0,7,14,…

   b. Define a table of 16b memory addresses; the addresses should be initialized to \$F000, \$F010, \$F020, …, \$F080

   c. Define an array of 8 uninitialized bytes

   d. Define an uninitialized array of LENGTH 16b values; assume LENGTH is already defined using an EQU pseudo-op.

5. Assume that ARRAY is an array of 8b signed integers. Write an instruction sequence that loads into accumulator A the value of ARRAY[4], using

   a. Extended addressing

   b. Indexed addressing (no offset)

   c. Offset indexed addressing, with ARRAY as the base

   d. Offset indexed addressing, with ARRAY as the offset

6. Suppose you wanted to write an assembly language sequence to initialize an 8 element array of 8b unsigned values to 0,1,2,...Assume the array has been defined somewhere else and is called DATA. If it is possible to write the sequence in loop form, then do so; otherwise, state that the loop form is not possible.

   a. Extended addressing

   b. Indexed addressing (no offset)

   c. Offset indexed addressing, with DATA as the base

   d. Offset indexed addressing, with DATA as the offset

7. Write a sequence of instructions that copies N bytes from ARRAY1 to ARRAY2. Use a loop. Assume N, ARRAY1 and ARRAY2 are defined elsewhere, and N is an 8b unsigned variable.

8. Using branch-to-else form, write a sequence of instructions that implements the code

   ```
   if P-Q > THRESHOLD then
           P=P-INCREMENT
   else
           Q=Q+INCREMENT
   ```

   Assume P and Q are unsigned 8b variables, THRESHOLD is a constant that has been defined using an EQU pseudo-op and INCREMENT is a constant variable defined with a DC.B pseudo-op.

9. Repeat the previous problem using the branch-to-then form.

10. Write code to compute R=|P|, where P is an signed 8b variable, using

    a. Branch-to then form.

    b. Branch to else form.

    Which form is better?

11. Write a sequence of instructions that determines the length of the NULL-terminated string at address STRING using

    a. While form

    b. Do-While form

    c. Which form is more appropriate in this case and why?

12. If a loop iterates at least once, what are the advantage(s) of using the Do-While loop form over the While loop form?

13. Write a sequence of instructions to implement the following switch without using a jump table. Assume variables and subroutines are defined elsewhere.

```
SWITCH (COMMAND)
CASE 'o':  OPEN()
      BREAK
CASE 's':  SAVE()
              BREAK
CASE 'q':  EXIT()
              BREAK
DEFAULT:   ERROR()
END SWITCH
```

14. Repeat the previous problem using a jump table.

15. Write a subroutine EXTEND(num) that returns in HX the sign-extended value of *num*, which is passed in accumulator A.

16. Provide a sequence of instructions to compute WN=EXTEND(N), where EXTEND() is as defined in the previous problem.

17. Provide a sketch of the stack frame for AND write a skeleton subroutine SKEL() that

    a. callee saves registers

    b. receives a single 16b parameter M via the stack

    c. returns a single 8b value in accumulator A

    d. has a 16b local variable VY initialized to M

    e. has an 8b local variable VX initialized to 0

    f. sets the return value to VX before returning

18. Provide a sequence of instructions to implement Z=SKEL(G), where SKEL() is as defined above. Assume Z and G are defined elsewhere.

19. Each of the following represents two methods to accomplish an operation. Describe what operation is being performed and indicate which is best in terms of (a) code size and (b) execution time

a.
```
CLRA                  CLR   $04
STA   $04
```

b.
```
AIS #2                PULA
                      PULA
```

c.
```
AIS #3                PULA
                      PULA
                      PULA
```

d.
```
CMP   #'c'            CBEQA #'c',TARGET
BEQ   TARGET
```

e.
```
BRCLR 0,$00,THERE     LDA #$01
                      BIT $00
                      BEQ THERE
```

20. Determine the static code size of the following sequence, assuming all variables are accessed via direct addressing.

```
          LDX   #13
WHILE     TSTX
          BEQ   ENDWHILE
          LDA   ,X
          ADD   SUM
          STA   SUM
          AIX   #-1
          BRA   WHILE
ENDWHILE  INC   SUM
```

21. Determine the execution time of the program in the previous problem, in

    a. Cycles

    b. Microseconds, assuming the CPU bus clock rate is 3.2 MHz.

22. Suppose you need to get the value N*X into accumulator A, where N is a constant. Two methods are shown below for when N=3. For what values of N is method a faster? Method b?

```
a.   LDA X    b.   LDA X
     ADD X         LDX #3
     ADD X         MUL
```

Biography

DOUGLAS H. SUMMERVILLE

Douglas H. Summerville is an Associate Professor in the Department of Electrical and Computer Engineering at the State University of New York at Binghamton. He was a Assistant Professor in the Department of Electrical Engineering at the University of Hawaii at Manoa and a visiting faculty researcher at the Air Force Research Laboratory, Rome, NY and the Naval Air Warfare Center, Patuxent River, MD. He received the B.E. Degree in Electrical Engineering in 1991 from the Cooper Union for the Advancement of Science and Art, and the M.S. and the Ph.D. degrees in Electrical Engineering from the State University of New York at Binghamton in 1994 and 1997, respectively. He has authored over 35 journal and conference papers. He is a senior member of the IEEE and a member of the ASEE. He is the recipient of one service excellence and two teaching excellence awards, all from the State University of New York. His research and teaching interests include microcontroller systems design, digital systems design and computer and network security. Email: dsummer@binghamton.edu.

Printed in the United States
by Baker & Taylor Publisher Services